Implementing Cellular IoT Solutions for Digital Transformation

Successfully develop, deploy, and maintain LTE and 5G enterprise IoT systems

Dennis McCain

BIRMINGHAM—MUMBAI

Implementing Cellular IoT Solutions for Digital Transformation

Group Product Manager: Rahul Nair
Publishing Product Manager: Surbhi Suman
Senior Content Development Editor: Adrija Mitra
Technical Editor: Rajat Sharma
Copy Editor: Safis Editing
Project Coordinator: Ashwin Kharwa
Proofreader: Safis Editing
Indexer: Sejal Dsilva
Production Designer: Shankar Kalbhor
Marketing Coordinator: Gaurav Christian

First published: February 2023

Production reference: 1100123

Published by Packt Publishing Ltd.
Livery Place
35 Livery Street
Birmingham
B3 2PB, UK.

978-1-80461-615-4

www.packtpub.com

*To my wife, Tamara, for her dedication and support to the family and me over the years.
To my son, Joshua, and daughter, McKenzie, believe in yourselves
you can achieve anything you set your mind to.*

– Dennis McCain

Foreword

I've had the opportunity to work with *Dennis McCain* on numerous IoT projects using the capabilities he describes in this book. Three of the most memorable projects will be revealed on these pages to give you a glimpse into Dennis's knowledge and experience.

The first was the development, launch, and support of a global IoT solution to track beverage coolers that are supposed to be inside buildings. IoT connectivity was important for tracking temperature, door opening and closing, and even the locations of the coolers because they could be stolen. This solution required integrating an IoT device with multiple cooler manufacturers and generations of cellular technologies to work worldwide, and almost every problem imaginable was encountered and solved along the way.

The second project was unraveling a tricky performance issue with a **Low-Power Wide-Area (LPWA)** solution that could be placed anywhere inside a building where there was power. Solving the problem required working with the customer, the cellular carrier, the device supplier, and the module and chipset supplier. This project was an example of how in-depth knowledge of the entire solution was necessary to achieve the end goal.

The third project was the development of a state-of-the-art smart label using an embedded SIM, LPWA module, and unique antenna and battery designs. Technical and performance issues, along with the constraints of size and cost, made this development challenging.

It's important to know that with IoT, new trails are blazed every day, and blazing new trails is hard work. You never know what's behind the next clump of trees. Dennis has been on point for much of this trailblazing. He knows the technology, the industry, the best practices, and the processes to navigate all aspects of IoT.

In this book, Dennis will be your guide to understanding the ins and outs of cellular IoT technologies as well as how to develop, deploy, and maintain IoT solutions. Even if you know wireless technology, IoT is a world of its own, and it's easy to get lost in the acronyms and the options. At the heart of IoT is the device itself, the gadget that connects whatever it is you want to monitor or control to an application. Dennis is an expert on IoT devices, so there's no one better at helping you understand the options available and identifying the right one for your use case. One of the biggest fears with IoT is the threat it could pose in the hands of bad actors, so Dennis will arm you with the right approaches to stay secure. And when you're ready to explore 5G and what it has to offer for massive IoT, enhanced mobile broadband, and ultra-reliable low-latency communication, Dennis can take you there.

Consider *Implementing Cellular IoT Solutions for Digital Transformation* as your guidebook. I wish you all the best on your IoT journey.

Cameron Coursey

Vice-President, Business Mobility & IoT Products, AT&T

Contributors

About the author

Dennis McCain is an IoT lead member of technical staff at AT&T and holds a master of science degree in electrical engineering from Texas A&M University. Dennis is the subject matter expert for IoT devices in the AT&T IoT business and has over 20 years of wireless technology experience with a focus on IoT, working in various roles in many markets, including utility smart energy, smart buildings, smart industrial, transportation and logistics, and smart homes.

I would like to thank my loving and patient wife for her continued support through the process of writing this book. I would also like to thank my IoT colleagues at AT&T for their encouragement in pursuing my goal of publishing this book.

About the reviewers

Anurag Kumar Singh has close to 12 years of experience designing, developing, and integrating various features and solutions in telecommunications systems, specifically in Access, Core and IoT. He has received a master's degree in computer science and provided POCs and training on end-to-end 4G, 5G, and cloud solutions for business clients of all sizes.

Innovation plays a huge role in networking, which is meant to connect people and technologies. I am thankful to all the people working toward telecommunication advancements over the last decade, with digital services reaching out to remote locations across the globe. Special mention to my family and friends for providing support and understanding for the time and commitment required to work this hard.

James Wolters has worked in a variety of technical leadership positions in the telecommunications industry over the past two decades. He has held network operations and network engineering roles with BellSouth Telecommunications and AT&T Mobility and has been focused on the IoT industry for the past 10 years. He holds a bachelor of science degree in aerospace engineering from the United States Naval Academy and is a graduate of the **Leaders for Global Operations (LGO)** fellowship program at the Massachusetts Institute of Technology.

Basavaraj Patil is a director for product development and realization at AT&T's IoT business unit. He leads a team of product architects and is responsible for the design, development, and deployment of end-to-end IoT solutions. His current focus is on 5G networks, LPWA, and various platforms within the IoT ecosystem. Basavaraj is based in Dallas, Texas, and works at the AT&T headquarters location downtown.

Basavaraj is a 25-plus-year veteran of the telecom/cellular/internet industry, starting in the early days of 2G and the mass adoption of mobile communications that has evolved to what we see today. He started at Bell Northern Research/Nortel Networks before moving to Nokia and, eventually, to AT&T.

His experience spans packet core networks, smartphones, data and voice, and internet protocols and technologies. He is the co-author of the book *IP in Wireless Networks*, published in 2003. He has also been actively involved in the **Internet Engineering Task Force (IETF)** and chaired several working groups, and co-authored a number of RFCs. He has also contributed to 3GPP, 3GPP2, and WiMAX standards. He holds multiple patents pertaining to cellular and IP-based communications.

Table of Contents

Preface   **xiii**

Part 1: Entering the World of the Internet of Things

1

Transforming to an IoT Business   3

Understanding IoT technologies	4	Improving the customer experience	9
Leveraging IoT for digital transformation	6	**Understanding IoT markets**	**9**
Monitoring and managing assets and inventory	6	Transportation	10
Increasing operational efficiencies and productivity	7	Supply chain logistics	10
		Industrial and manufacturing	14
Creating smart factories	7	Healthcare	15
New business opportunities	8	Energy/utilities	16
		Summary	**17**

2

Understanding IoT Devices and Architectures   19

High-level IoT architecture	**20**	The UE	26
The link layer	21	eNodeB	27
The network layer	22	EPC	27
The transport layer	22	**An overview of IoT device types and use cases**	**28**
The application layer	22		
The data format	24	Serial modems	29
The cellular IoT network architecture	**25**	Routers	29

Gateways   30
Remote monitoring devices   30
Asset trackers   31

Best practices   **32**
Leveraging off-the-shelf
technology components   32

Leveraging established IoT technology partners 33
Planning on solution life cycle management   33
Slow and steady wins the race   33
Security is paramount   33

Summary   **34**

3

Introducing IoT Wireless Technologies   35

Licensed versus unlicensed
technologies   35
Network topologies   37
WWAN and WLAN IoT technologies   37
Bluetooth/BLE   38
Wi-Fi   38
LoRa and LoRaWAN   39
LR-WPAN (IEEE 802.15.4)   40

Technical and cost comparisons   41
Frequency spectrum   42
Range   43
Power consumption   43
Data throughput   44

Module cost   44

Wireless technology use cases   44
Bluetooth/BLE   44
Wi-Fi   45
LoRa   46
LR-WPAN   46

Best practices in deploying IoT
wireless technologies   47
1. Consider the network coverage required   47
2. Consider the long-term network
maintenance/support required   47
3. Leverage existing network infrastructure   47

Summary   48

Part 2: Deep Dive into Cellular IoT Solutions

4

Leveraging Cellular IoT Technologies   51

The evolution of cellular technology   52
LTE technologies   54
LTE network   54
LTE radio technology   59
LTE technology ecosystem   63
LTE IoT use cases   64

LTE LPWA design and best practices   65
PSM and eDRX low-power features   67
LTE LPWA best practices   68

LTE LPWA use cases   70
Summary   72

5

Validating 5G with IoT   73

Overview of 5G   74
5G network architecture   74
5G radio technology   75
5G frequency spectrum   76
5G network slicing   77

5G and IoT   78
5G NR RedCap   78

5G use cases   79
Industrial automation (Industry 4.0)   80

Remote surgery   80
Smart vehicles (autonomous vehicles and V2X)   80
VR/AR   80

Best practices in 5G solutions   81
5G network carriers   81
5G private networks   81
5G IoT devices   82

Summary   82

6

Reviewing Cellular IoT Devices with Use Cases   83

Cellular IoT device architecture   83
Processing unit   84
Cellular module   85
SIM   86
RF   87
Power unit   87
Sensors   88
Input/output   89
WLAN radios   90

Cellular IoT device types   91
Alarm panel   92
Camera   92
Computer – in vehicle   92
Emergency phone   92
Hotspot   92
Lighting   93
Medical telematics   93
Metering   93
Modem – embedded   93

mPERS   94
POS   94
Remote control device   94
Rugged handheld   94
Sensors   94
Smart home   95
Tracking   95
Vehicle telematics control unit   95
Vehicle OBD II   95
Vending telemetry   95
Wearable   96

Cellular IoT device carrier certifications   96
Edge computing   97
Best practices in cellular IoT devices   99
Buying versus building   99
Carrier certification   99
Device management   100

Summary   100

7

Securing the Internet of Things   101

An overview of the IoT security ecosystem	102	IoT devices	110
IoT security challenges	103	IoT device hardware security	113
Threats to IoT devices and solutions	104	IoT device software security	113
		IoT device OS security	114
Proposed IoT security framework	105	IoT device connectivity interface security	115
Endpoint layer	106	IoT device network connectivity security	115
Network layer	107	IoT device Wi-Fi security	116
Application layer	110	Summary	117
Best practices for securing			

8

Implementing an IoT Solution with Case Studies   119

Exploring the IoT solution business cases	119	Connected Cooler	130
		Smart Label	132
Connected Cooler	120	Lessons learned	135
Smart Label	121	Piloting the solution	135
Investigating the IoT device architectures	123	Device management	136
		Life cycle management	136
Connected Cooler	124	Summary	138
Smart Label	127		
Analyzing the network architectures	129		

Part 3: Cellular IoT Solution Life Cycle and Future Trends

9

Managing the Cellular IoT Solution Life Cycle 141

The challenges of IoT life cycle management	**141**	Monitoring and management	146
		Sunset	147
IoT devices	142		
IoT wireless services	142	**Best practices**	**147**
IoT platforms	143	The designing and planning stage	148
		Provisioning	149
The cellular IoT life cycle	**144**	Deployment	150
Designing and planning	144	Monitoring and management	151
Provisioning	145	**Summary**	**152**
Deployment	146		

10

Looking at the Road Ahead 153

Emerging cellular IoT technologies	**154**	The IoT for sustainability	161
Private LTE and 5G CBRS networks	154	New IoT business models	161
iUICC and eUICC SIM technologies	156	**Getting started on your IoT journey**	**163**
IoT SAFE	157	**Best practices in developing and launching your IoT solution**	**165**
IoT trends and business models	**157**		
Edge computing	158	Leveraging the IoT partner ecosystem	165
AI and ML	158	Focus on interoperability between IoT systems	166
Interoperability of IoT solutions	159	Planning on an 18-24 month launch schedule	166
A social and ethical IoT	160	**Summary**	**166**

Index 169

Other Books You May Enjoy 180

Preface

The IoT market, especially in the area of cellular IoT solutions using LTE and 5G technologies, has grown exponentially in the past 5 years as businesses are realizing how IoT is a critical enabler for revenue growth, with new business models and operational cost savings. As IoT is either in the planning or development phase for many businesses, it is important to understand end-to-end IoT technologies and industry best practices to create optimal IoT solutions. While the enterprise information technology team typically responsible for implementing IoT solutions may understand **wireless local area network (WLAN)** technologies such as Wi-Fi, cellular wireless wide area network technologies introduce a number of new, technical challenges that need to be understood before implementing an enterprise IoT solution. This book endeavors to provide a holistic review of the primary IoT wireless technologies, devices, and architectures with a focus on cellular IoT solutions.

Both developers and engineering managers planning an IoT solution will be able to put their knowledge to work with this practical guide to understand the fundamentals of IoT. This book provides an understandable overview of the IoT wireless technologies, especially cellular, with real-world case studies and best practices to follow at each phase of the IoT solution life cycle. You will learn about the IoT market and the benefits of an IoT solution followed by a review of the various IoT wireless technologies, with a focus on cellular LTE and 5G. You will then learn about the main components and best practices for a successful end-to-end cellular IoT solution, followed by practical case studies of real-world cellular IoT solutions. This book will take you from an understanding of IoT wireless technologies to the business impetus and guide for a complete IoT solution implementation, concluding with real-world case studies and future IoT trends. By the end of this book, you will be able to identify the best wireless technologies for IoT use cases and will understand the architecture and best practices for a successful cellular IoT solution addressing the key pain points in the solution life cycle.

Who this book is for

This book is for IoT technology managers, leaders, C-suite executives, and decision makers considering or currently developing IoT solutions based on cellular technologies such as LTE and 5G. We assume the reader understands the importance of IoT connectivity in the context of their IoT solution.

What this book covers

Chapter 1, *Transforming to an IoT Business*, provides an introduction to the Internet of Things, the top IoT markets, and why it is important for a business to consider an IoT solution. It introduces the basic technologies and architecture of an IoT solution and what types of "things" are connected and why.

Chapter 2, Understanding IoT Devices and Architectures, provides an introduction to IoT solutions, from the architecture of the IoT device to cloud applications, with descriptions of each component and best practices in designing an enterprise IoT solution.

Chapter 3, Introducing IoT Wireless Technologies, provides an overview of the primary IoT wireless technologies, including both licensed and unlicensed WWAN and WLAN technologies. This chapter provides a comparison of cellular and other wireless technologies deployed on the market with a review of use cases, along with technical and cost trade-offs.

Chapter 4, Leveraging Cellular IoT Technologies, goes into detail on cellular wireless technologies, especially **Low Power Wide Area (LPWA)** technologies. It provides an overview of LTE wireless technologies, best practices in using these technologies, and the new IoT use cases enabled by these technologies.

Chapter 5, Validating 5G with IoT, goes into detail about the newest cellular technology, 5G, which has unique features enabling new high-bandwidth, low-latency IoT applications. The chapter provides an overview of 5G with use cases and best practices in using 5G technologies in IoT solutions.

Chapter 6, Reviewing Cellular IoT Devices with Use Cases, builds on the high-level overview of common IoT devices and use cases. It provides more detail on cellular IoT device types, use cases, and carrier certifications, including a review of edge computing and machine learning. The chapter concludes with a review of best practices in selecting cellular IoT devices for an enterprise IoT solution.

Chapter 7, Securing the Internet of Things, looks at privacy and security at all levels of the IoT solution architecture, including the device and connectivity – a critical part of any large-scale IoT solution. The chapter provides details on critical privacy and security guidelines and best practices for a robust, secure end-to-end IoT solution.

Chapter 8, Implementing an IoT Solution with Case Studies, crystallizes the review of complete IoT solutions from business case to technical execution. This chapter presents two real-world cellular IoT solution implementations covering the technical decisions for each component of the IoT solution presented in the previous chapters.

Chapter 9, Managing the Cellular IoT Solution Life Cycle, presents the full cellular IoT solution life cycle, describing the challenges and best practices at each stage to help ensure the success of an enterprise IoT solution.

Chapter 10, Looking at the Road Ahead, presents some emerging cellular IoT technologies as well as new IoT trends and business models likely to emerge in the next 10 years. The chapter provides insight into and best practices for getting started with designing and implementing an enterprise IoT solution.

To get the most out of this book

This book is intended for readers that just want to know more about all aspects of IoT solutions or readers that are considering implementing an IoT solution and realize the importance of IoT wireless connectivity.

Download the color images

We also provide a PDF file that has color images of the screenshots and diagrams used in this book. You can download it here: `https://packt.link/NOvWZ`.

Conventions used

There are a number of text conventions used throughout this book.

Bold: Indicates a new term, an important word, or words that you see onscreen. Here is an example: "The first case study is the **Connected Cooler** solution, and the second case study is the **Smart Label** solution."

Italics: Indicates references to another chapter, another section in the same chapter, or a particular image in a chapter. This style is also used to highlight keyboard keys. Here is an example: "These network architectures are shown in *Figure 5.2*."

> **Tips or important notes**
> Appear like this.

Get in touch

Feedback from our readers is always welcome.

General feedback: If you have questions about any aspect of this book, email us at `customercare@packtpub.com` and mention the book title in the subject of your message.

Errata: Although we have taken every care to ensure the accuracy of our content, mistakes do happen. If you have found a mistake in this book, we would be grateful if you would report this to us. Please visit `www.packtpub.com/support/errata` and fill in the form.

Piracy: If you come across any illegal copies of our works in any form on the internet, we would be grateful if you would provide us with the location address or website name. Please contact us at `copyright@packt.com` with a link to the material.

If you are interested in becoming an author: If there is a topic that you have expertise in and you are interested in either writing or contributing to a book, please visit `authors.packtpub.com`.

Share Your Thoughts

Once you've read *Implementing Cellular IoT Solutions for Digital Transformation*, we'd love to hear your thoughts! Scan the QR code below to go straight to the Amazon review page for this book and share your feedback.

https://packt.link/r/180461615X

Your review is important to us and the tech community and will help us make sure we're delivering excellent quality content.

Download a free PDF copy of this book

Thanks for purchasing this book!

Do you like to read on the go but are unable to carry your print books everywhere? Is your eBook purchase not compatible with the device of your choice?

Don't worry, now with every Packt book you get a DRM-free PDF version of that book at no cost.

Read anywhere, any place, on any device. Search, copy, and paste code from your favorite technical books directly into your application.

The perks don't stop there, you can get exclusive access to discounts, newsletters, and great free content in your inbox daily

Follow these simple steps to get the benefits:

1. Scan the QR code or visit the link below

https://packt.link/free-ebook/9781804616154

2. Submit your proof of purchase
3. That's it! We'll send your free PDF and other benefits to your email directly

Part 1:
Entering the World of the Internet of Things

To begin our IoT journey, we will start with an overview of the top IoT markets and underlying IoT devices, technologies, and architectures that enable enterprise IoT solutions. This will set the foundation for our deep dive into cellular IoT solutions in *Part 2*.

This partcontains the following chapters:

- *Chapter 1, Transforming to an IoT Business*
- *Chapter 2, Understanding IoT Devices and Architectures*
- *Chapter 3, Introducing IoT Wireless Technologies*

1
Transforming to an IoT Business

The majority of people are familiar with the **Internet of Things** (**IoT**), which was first coined in 1985 and was born according to Cisco Systems in 2008 when more things were connected to the internet than people. In practical terms, IoT is defined as physical objects that connect and exchange data with other devices and systems over the internet. Now, IoT is ubiquitous, especially in the consumer market, where home automation/monitoring is now quite common. As of 2020, there were over 12 billion connected devices with a forecast of more than 30.9 billion IoT devices worldwide by 2025 (source: *Business Insider, IoT Analytics, Gartner, Intel, Statista*). There were 5.8 billion connected automotive and enterprise IoT devices by the end of 2020, and it is expected that more than 15 billion enterprise IoT devices will be connected by 2029 (source: *Gartner*), which is why enterprise IoT is the focus of this book. While over 98% of business leaders have an understanding of IoT, statistics show that many are unclear of the exact definition of the term (source: *Fierce Electronics*), which undermines the full potential of enterprise IoT solutions and is the reason we decided to write this book. The enterprise market is where IoT is making the biggest impact in terms of digital transformation, with operational cost savings and new business models. By the end of this chapter, you will have a good understanding of the underlying IoT solution technologies and markets, as well as how IoT solutions can be used to transform a business with operational cost savings, improved efficiencies, and new business models.

In this chapter, we will cover the following main topics:

- Understanding IoT technologies
- Leveraging IoT for digital transformation
- Discovering the top IoT markets

Understanding IoT technologies

In this section, you will learn about the underlying IoT technologies while using a high-level IoT solution end-to-end architecture as a guide. This includes IoT devices, connectivity, data analytics, and applications. The goal of this section is to provide you with a practical understanding of an IoT solution that will form the basis for an enterprise's digital transformation, as discussed in the next section.

At a high level, an IoT solution consists of four basic layers, as shown in the following figure:

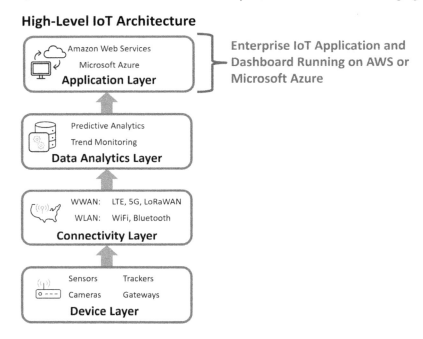

Figure 1.1 – High-level IoT architecture

The **device layer** is the actual physical device or *thing* in IoT that is connected to the internet. It captures critical IoT sensor data such as temperature, humidity, light, and air quality for remote monitoring applications. We will explore IoT devices in more detail in *Chapter 6, Reviewing Cellular IoT Devices with Use Cases*, but some common IoT devices that gather this sensor data include gateways, routers, and asset trackers. This sensor data can include location data for assets such as cars and enterprise fleets, health data for **remote patient monitoring** (**RPM**), or video data for security. We will cover IoT device types in more detail in *Chapter 2, Understanding IoT Devices and Architectures*, and describe how an IoT device is the foundation for an IoT solution. We will also discuss the near real-time processing of data at the device layer, known as **edge computing**, which is a growing and important trend that is driving tremendous growth in enterprise IoT.

The **connectivity layer** is the wireless or wired connectivity, which is the *gateway* to the internet. This could be a **wireless local area network** (**WLAN**) technology such as Wi-Fi, Bluetooth, or Zigbee or a

wireless wide area network (**WWAN**) technology such as cellular or LoRaWAN. As we will discuss in *Chapter 2, Understanding IoT Devices and Architectures*, and *Chapter 3, Introducing IoT Wireless Technologies*, WLAN technologies rely on a WAN gateway device to backhaul connectivity to the internet. Although several wireless technologies enable IoT, which we will present in *Chapter 3, Introducing IoT Wireless Technologies*, we will make the case that cellular technologies such as LTE and 5G are unique enablers for the new and transformative IoT business models, which is part of the reason why cellular IoT devices grew 18% year-over-year to reach 2 billion by the end of 2021 (source: *IoT Analytics*). More specifically, **Low Power Wide Area** (**LPWA**) cellular technologies such as **LTE Category M** (**LTE-M**) and **Narrow Band IoT** (**NB-IoT**) offer unique features in terms of lower cost and power that further enable new IoT applications. In *Chapter 3, Introducing IoT Wireless Technologies*, we will provide much more detail on both licensed and unlicensed wireless IoT technologies, especially LPWA, and how they enable the growth of enterprise IoT solutions and new IoT business models. In *Part 2*, we will do a deep dive into cellular IoT wireless technologies.

The **data analytics layer** is where the IoT device data is processed and is typically implemented in combination with the **application layer** in the domain of a **Cloud Service Provider** (**CSP**) such as **Amazon Web Services** (**AWS**) or **Microsoft Azure**. The data analytics layer is where the value of the IoT device data is realized in terms of data analytics and actionable data for an enterprise IoT solution. As mentioned earlier, for many IoT applications, much of the IoT data processing is moving from the data analytics layer in the cloud to the device, which is known as "edge" data processing. This enables more real-time and low-latency applications such as intelligent transportation (for example, autonomous vehicles), augmented/virtual reality, and video intelligence. We will go into more detail on IoT data processing in the context of IoT architectures in *Chapter 2, Understanding IoT Devices and Architectures.*

The **application layer** is the dashboard for the IoT solution that pulls together the IoT device data with the connectivity and processing into a meaningful presentation. For example, in the case of asset tracking, this would be a map showing the location and contiguous path of an asset such as a fleet vehicle or container, along with temperature and humidity sensing data/alerts along the way. Behind the application is where data processing takes place. This involves the time series data from the IoT device along with sensor and location data. In this example, the asset tracking application in combination with the underlying data processing provides not only a better user experience in monitoring the asset but also identifies patterns that drive business decisions around real operational cost savings. We will cover the IoT application protocols, including MQTT and CoAP, in more detail in *Chapter 2, Understanding IoT Devices and Architectures.*

We will review the end-to-end IoT solution architecture in more detail in *Chapter 2, Understanding IoT Devices and Architectures*, but at a high level, these four IoT layers are the basis for an enterprise IoT digital transformation in terms of improved customer experience, real-time insights leading to operational cost savings, and new service business lines. In the *Understanding IoT markets* section, we will discuss the top IoT markets, along with some example applications in each market, that helped create a digital transformation strategy that is being adopted by more and more businesses. While in 2018, 57% of businesses adopted IoT in some way, this increased to 94% in 2021 (source: *Aruba*

Research Report, Microsoft). Moreover, 83% of organizations that employed IoT technology have reported a significant increase in business efficiency (source: *Aruba Research Report, Microsoft*). In the next section, we will learn how IoT can drive enterprise digital transformation.

Leveraging IoT for digital transformation

What is the role of IoT in enterprise digital transformation? In short, IoT solutions act as a connection between the physical assets of a business and the **information technology** (**IT**) infrastructure, leading to improved operational efficiencies as well as new and better customer experiences. To provide the impetus for implementing an IoT solution in your business, we will describe five areas where IoT can enable digital transformation in your business:

1. Monitoring and managing assets and inventory.

2. Increasing operational efficiencies and productivity.

3. Creating smart factories.

4. Driving new business models.

5. Improving the customer experience.

Let's start by reviewing one of the most common areas where an enterprise IoT solution can transform your business – monitoring and managing assets and inventory.

Monitoring and managing assets and inventory

A critical area where an IoT solution can transform your business is providing information on the health and location of various business assets, shipments, and stock inventory:

| Outdoor Monitoring | Fleet Management | Livestock Monitoring |
| | Air Quality | Surveillance |

| Product Integrity | Cold Chain Logistics Produce Tracing | Pharmaceuticals |

| Indoor Location | Warehouse Inventory | Mobile Assets |
| | Worker Safety | |

| Shipment Monitoring | Supply Chain Logistics | |
| | Disposable Trackers | Container Tracking |

Figure 1.2 – Asset tracking and monitoring use cases

One of the leading areas where IoT can transform a business is asset monitoring and inventory management. In terms of asset monitoring, an IoT solution can provide automatic notifications on the health and location of business assets such as fleet vehicles/trailers, containers, pallets/bins, retail coolers, and machinery. This enables a business to reduce human error and improve daily operational efficiencies and asset downtime due to maintenance, also known as predictive maintenance. Business assets are the lifeblood of a successful business, so effectively monitoring and managing these assets is a fairly easy way to realize significant operational cost savings. An IoT solution is also ideal for inventory management, where the location, temperature, and stock levels of inventory are critical to the success of a business supply chain. An IoT solution with wireless technologies such as RFID, Bluetooth, Wi-Fi, and cellular in conjunction with the appropriate IoT devices and applications virtually eliminates the outdated manual processes associated with inventory management, which are prone to human error and theft, which increase the operational cost of a business. Along the lines of asset management is shipment tracking with IoT-enabled asset monitoring of not only location but also temperature, humidity, and tampering to directly validate the integrity of shipments. This will be discussed in more detail in the next section on IoT markets.

An IoT solution can further transform your business by improving overall operational efficiency and productivity through the seamless integration of business processes and systems.

Increasing operational efficiencies and productivity

There are many ways an IoT solution can increase efficiency and productivity in a business. In general, an enterprise IoT solution can efficiently tie inventory management systems and factory automation/monitoring with **customer relationship management** (CRM) systems and logistics for improved customer relationships and operational cost savings. In a factory setting, the data from IoT devices can provide unique insight into the production line processes and shipping delivery times, which helps operations complete quicker and more cost-effectively. Moreover, enterprise IoT solutions can reduce employee workloads and allow for further automation, creating cost-saving efficiencies. **Industrial IoT (IIoT)** solutions are revolutionizing the manufacturing industry by monitoring machine performance and conditions for predictive maintenance to eliminate production line bottlenecks due to machine downtime.

This leads to our third related area where IoT is transforming businesses, which is the Smart Factory.

Creating smart factories

IoT has been the enabler for factory automation and predictive analytics, which has been called **smart factory**:

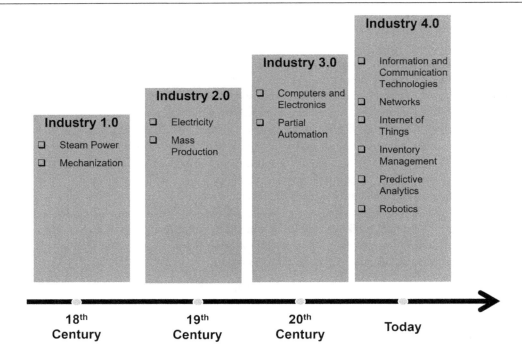

Figure 1.3 – Evolution of Industry 4.0 and smart factories

As discussed earlier, IIoT solutions can create several operational efficiencies in a factory setting, which, together with the IoT data analytics from staff, supply chain, stock inventory, and machine health, allows for what has been called the **4th industrial revolution** or **Industry 4.0**. The hallmarks of the *smart factory* in Industry 4.0 that have been enabled by IoT are predictive analytics for machine maintenance, inventory management, and smart production workflows that are more automated (for example, industrial robotics) for improved operational and cost efficiencies, as well as reduced waste. With a smart factory, potential disruptions are reduced, and production is both more efficient and predictable.

With the adoption of IoT, businesses are now also able to create entirely new business models, thus creating new revenue streams and providing added value to customers.

New business opportunities

Businesses that use IoT solutions have created entirely new service business models, which allow them to track and monitor the performance of their products over their full life cycle, enabling new value-added services for their customers. For example, in the healthcare market, IoT enables both RPM and telemedicine services, which we will cover later in this chapter. In the agricultural market, IoT enables crops as a service with full crop life cycle management from seed to shipment. An IoT supply chain logistics solution provides not only real-time location data but also data on the temperature, humidity, and integrity of shipped assets such as perishables, including pharmaceuticals and food, which provides significant value to end customers and consumers. These value-added services create a closer relationship and trust with their customers, enabling future business.

By implementing IoT solutions, businesses can now offer subscription services around these IoT-enabled value-added services that avoid the upfront capital expense associated with traditional products. With a continuous connection to customers through an IoT solution, businesses can now develop recurring revenue *as-a-service* models where the customer pays for continuous value. For example, several new IoT-enabled businesses offer fleet/asset management applications that provide a subscription-based offering to their customers to track and manage their fleet and other critical assets without the upfront capital expense for the devices or infrastructure. The *as-a-service* IoT solution model, which uses advanced data analytics, can be extended to many other subscription-based models, such as security monitoring as a service, health monitoring as a service, crops as a service, and predictive maintenance as a service.

Ultimately, IoT solutions not only benefit businesses but also customers with improved, more personalized customer engagements throughout the product life cycle.

Improving the customer experience

There are many ways IoT solutions improve customer experiences. IoT solutions allow businesses to collect data at various points in a product life cycle from the time it is produced to the time it is retired, providing deep insight into how the product is used to better personalize the customer experience. This improves customer loyalty and trust in the brand. For example, we are probably all familiar with the ease of making purchases at retail stores and kiosks using smartphone apps and smart point-of-sale IoT devices, which improves and personalizes the customer experience with that business. Retailers are also using IoT to manage shelf stocks and monitor shopper behavior to better serve in-store customers.

Now that you have some insight into how IoT can enable a digital transformation in your business, let's review the leading IoT markets with some example IoT applications in each before considering some common business problems solved by implementing an IoT solution.

Understanding IoT markets

As of this writing, the top five enterprise IoT market segments are as follows:

1. Transportation
2. Supply chain logistics
3. Industrial and manufacturing
4. Healthcare
5. Energy/utilities

In this section, we will review and define each of these segments and provide insights into how IoT is transforming these markets with some example IoT applications.

Transportation

Transportation includes both automotive *connected cars* and enterprise logistics, which depend on air, rail, and truck transportation. As we will discuss in the next section on the logistics market, IoT asset monitoring solutions in transportation are a critical part of improving efficiencies in supply-chain logistics. Concerning connected cars, Cisco estimated that connected vehicle applications will be the fastest-growing IoT market with a 30% compound annual growth rate (source: *Cisco 2020*). The number of connected cars is projected to be over 400 million by 2025 (source: *Statista 2021*). In terms of fleet management, IoT solutions enable an enterprise to better manage fuel consumption, optimize routes, and maintain its fleet to reduce operational expenses and increase productivity. Fleet management IoT solutions also improve fleet safety and compliance with the new **electronic logging device** (**ELD**) government mandates, which further improves overall fleet safety. A typical IoT solution for fleet management would include a vehicle location tracking device with the capability of monitoring vehicle maintenance and engine runtimes. Typically, there is also a gateway (ELD) device to automatically log drive times, which avoids the human error of manual drive logs. Many fleet management solutions now also include both security cameras and dash cameras to monitor both the driver's behavior and the behavior of other drivers on the road for potential accident investigations and security. Almost all key components of a fleet, including location, vehicle behavior, engine maintenance/performance, idle time, cargo status, driver behavior, and security events, can be monitored with an IoT solution.

Nearly all new cars have integrated telematic IoT connectivity, which enables not only in-vehicle infotainment applications but also embedded communications, vehicle updates, and predictive maintenance, which improves the customer experience and provides new service business models for automotive OEMs. This IoT connectivity has been extended further to include autonomous driving and vehicle communication with other vehicles, city infrastructure, pedestrians, and cloud applications, which has been termed **Cellular V2X** (**C-V2X**). As we will discuss in *Chapter 5*, *Validating 5G with IoT*, low-latency, high-bandwidth 5G cellular technologies play a critical role in enabling these new connected car applications. Also included in the connected car IoT market are several IoT use cases based on the installation of after-market IoT gateways or dash cameras in the vehicle. For example, several insurance companies now offer insurance discounts to drivers with good driving behavior who use IoT sensor devices in their vehicles that detect speed, hard braking, and hard acceleration. Much like the fleet management IoT applications described earlier, dash camera applications can improve the safety and security of drivers.

Supply chain logistics

As shown in *Figure 1.4*, supply chain logistics is the networked infrastructure of suppliers, transportation, production, warehousing, and stock inventory that provides the end-to-end connection between suppliers and wholesale/retail customers:

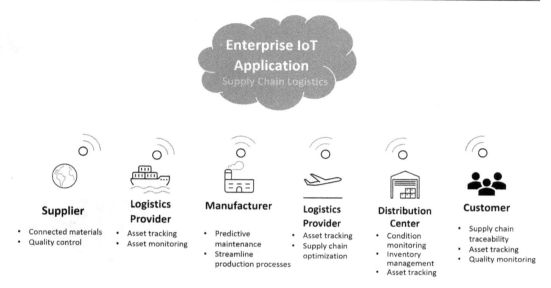

Figure 1.4 – Supply chain logistics IoT use cases

The importance of supply chain logistics to the global economy was made clear with the recent pandemic in which shortages in semiconductor chips and raw materials have impacted all areas of the global economy, especially manufacturing. As shown in the preceding figure, with IoT technologies, it is now possible to have end-to-end visibility of the entire supply chain from the raw materials to the end customer. The supply chain disruptions because of the pandemic have been the impetus for change in supply chain operations worldwide and have increased the focus on IoT solutions to increase efficiencies and overall visibility. The global supply chain management market size was valued at USD 16.64 billion in 2021 and is expected to expand at a CAGR of 10.8% from 2022 to 2028 (source: *Grand View Research 2022*). There are three key areas of supply chain logistics where IoT solutions can have the most impact. These are as follows:

- Inventory/warehouse management
- Asset monitoring
- Transportation and fleet management

Let's go into more detail on the IoT solutions deployed in these areas of the supply chain.

Inventory/warehouse management

The most common application of IoT in supply chain logistics is in the area of inventory and warehouse management, where IoT solutions can not only track incoming and outgoing shipments but also manage inventory levels by working with enterprise ordering and inventory management systems:

Figure 1.5 – IoT solutions in inventory/warehouse management

One of the fundamental needs in supply chain logistics is being able to locate and manage warehouse inventory in near real-time, which improves the visibility/traceability of stock and provides customers with more predictable orders and lead times. Before the advent of IoT solutions, warehouse inventory management was a mostly manual process requiring lots of people to locate, track, and report inventory coming in and out of a warehouse or distribution center.

With an IoT solution, several wireless and device technologies significantly improve the visibility and management of warehouse inventory. Today, cellular or Wi-Fi-connected barcode scanners are commonplace to track inventory, but there are also RFID and Bluetooth tags attached to assets that can automatically provide both the location and status (for example, temperature) of inventory within the warehouse. As part of an IoT solution, RFID readers and Bluetooth location beacons are placed strategically in warehouse and distribution centers to automatically track/monitor inventory along the supply chain and report back to the enterprise IoT platform. This not only increases the visibility of the inventory but also reduces the errors associated with manual processes and enables operational cost savings with reduced workloads.

Asset monitoring

In the context of a complete supply chain logistics IoT solution, it is important to not only manage/ monitor inventory and material in warehouses but also along the entire supply chain from the factory to the end customer, as shown in the following figure:

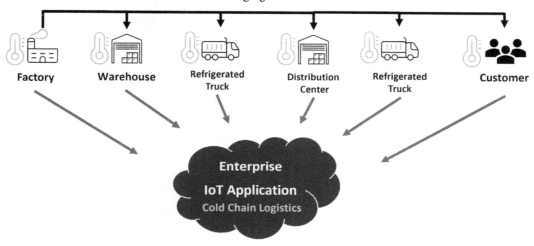

Figure 1.6 – IoT solutions in supply chain asset monitoring

Especially with high-value assets, products, and pharmaceuticals, asset monitoring along the supply chain is an important IoT use case that is becoming more common as the cost of wireless monitoring devices with sensors are decreasing. As part of an IoT solution, a low-cost/disposable wireless monitoring device is attached to the target asset, and it reports to an IoT platform either periodically or based on events (for example, movement, sensor threshold exceeded, tamper, and shock). This not only provides visibility along the supply chain but also provides traceability for issues on the journey (for example, theft) and validates the integrity of the asset from the source to the destination. In some IoT solutions, the trace of an asset's journey and everything that has happened to it along the way are sent to private blockchain networks as an immutable record of the asset's integrity.

Transportation and fleet management

Transportation and fleet management is the third area of supply chain logistics where an IoT solution can have a significant impact in terms of operational cost savings and asset integrity. As shown in the following figure, fleet management IoT solutions can provide data on both the cargo location and environment, as well as the vehicle's health:

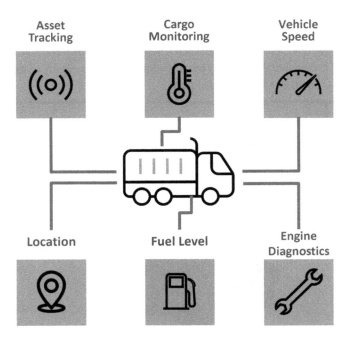

Figure 1.7 – IoT solutions in transportation/fleet management

As part of an IoT fleet management solution, IoT provides not only the shipment location and environmental conditions but also the vehicle condition, which combines fleet management with supply chain visibility/monitoring. Domestically, this fleet is generally trucks and trailers, which would have integrated IoT devices for cargo tracking and vehicle monitoring, as shown in the preceding figure, but there are several modes of transportation in supply chains including air, sea, and rail where supply chain visibility is important. The challenge is to provide shipment visibility and status across all modes and carriers globally. A global IoT solution using a global WWAN cellular technology in combination with an appropriate IoT shipment monitoring device with data logging can address this challenge by providing the traceability and asset integrity validation discussed earlier. This use case will be discussed in more detail in a case study in *Chapter 8, Implementing an IoT Solution with Case Studies*, where we will cover global cellular technologies and devices as part of a supply chain logistics IoT solution.

Industrial and manufacturing

Earlier, we discussed how IIoT and Industry 4.0 are transforming the manufacturing industry by improving workflow efficiencies and factory automation. Manufacturing is the sector most affected by IoT, with a potential economic impact of $3.9 trillion by 2025 (source: *McKinsey Global Institute*). According to GE, 58% of manufacturers say IoT is required to digitally transform industrial operations (source: *GE*), with 80% of industrial manufacturing companies having adopted IoT in some way (source: *Security Today*).

There are three main reasons why production/manufacturing is one of the leading IoT markets. First, IoT solutions enable production machine monitoring and predictive maintenance, even in older factories without significant automation saving downtime. Retrofitted IoT remote connectivity and sensors on machines in a factory allow operators to quickly identify issues and implement predictive maintenance on the machine, minimizing downtime and significantly improving production efficiency. Moreover, machine learning and artificial intelligence in the IoT devices connected to machines can identify anomalous patterns and alert an operator to act. As an example, IoT vibration monitoring of a machine can be used to identify unusual behavior such as an impending machine bearing failure before it is a costly machine failure and production stopper. Second, IoT enables remote production control of machines, where operators can remotely fix/tune many performance issues without hindering the production flow. In this case, a single operator can monitor and control several machines, which reduces the factory workload and enables operational cost savings. With low-latency 5G technologies in the IoT solution, this monitoring and control can be near real-time, increasing operational efficiency. Finally, IoT enables industrial robotics, which reduces factory workloads and increases efficiency at all stages of manufacturing, and with 5G technologies in the IoT solution, there is more flexibility and mobility of production lines where robots can be placed on the factory floor with safer human-robot interactions.

Healthcare

With the recent pandemic and the increased need for RPM, IoT solutions have been a focus in healthcare. Implementing IoT RPM solutions can offload our critical healthcare infrastructure and foster better patient care:

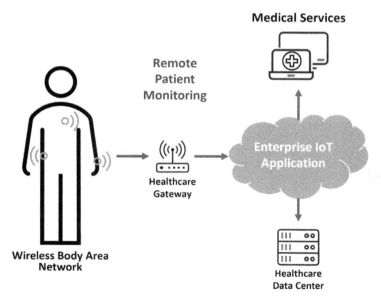

Figure 1.8 – IoT solutions in healthcare RPM

IoT in healthcare has been termed the **Internet of Medical Things (IoMT)** and includes RPM-connected devices and applications, as well as wearables. As shown in *Figure 1.8*, with RPM, wireless medical devices such as blood pressure monitors, glucose monitors, oximeters, and ECGs can be connected wirelessly typically through an IoT gateway to the internet for evaluation by a medical professional. RPM with IoMT lends itself to many healthcare "as-a-service" telemedicine models, and with low-latency and highly reliable 5G technologies, remote and robotic-assisted surgeries have been proven to work well, which overcomes geographic limitations for surgical procedures in remote areas. The importance of RPM and care was clear with the recent COVID-19 pandemic, where care facilities and hospitals were overwhelmed with patients. Certainly, IoT-enabled RPM with patients at home would have reduced the strain on our healthcare infrastructure. With the aging population and chronic care monitoring becoming more prevalent, IoMT services will become more important in the next 5 years. Deloitte says that the IoMT market will reach $158.1 billion in 2022, and Goldman Sachs claims that healthcare organizations save $300 billion annually from RPM and other technological benefits.

Energy/utilities

Much like IIoT, IoT solutions are transforming our utility infrastructure from generation and storage to transmission, distribution, and consumption, as shown in the following figure:

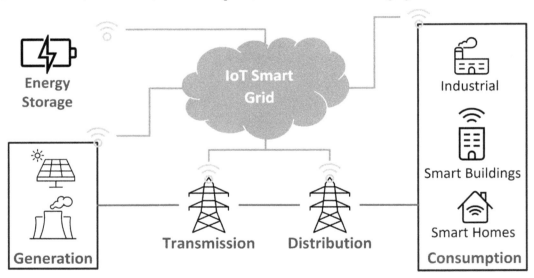

Figure 1.9 – IoT solutions in the Smart Grid

IoT in energy and utilities has been termed the **Smart Grid**. With IoT-enabled remote monitoring and control of the energy grid infrastructure, the overall network efficiency and resiliency from generation to transmission and distribution and ultimately businesses and homes can be optimized. Some Smart Grid IoT applications include connected electric meters, real-time transmission/distribution outage notifications and restorations, remote monitoring/control of power generation and storage, and dynamic load distribution. Especially with aging utility infrastructure and new forms of energy generation, storage, and distribution, the Smart Grid will become even more critical in the next 5 years. According to Future Market Insights in 2022, IoT in the utility market is extrapolated to reach a value of $129.1 billion by 2032. The global focus on dramatically reducing carbon emissions is another strong driver for this growth in the Smart Grid. Government mandates in carbon emission reductions from power generation plants drive the need for constant monitoring enabled by IoT.

In all the markets where IoT is deployed, the huge amount of data garnered from these deployments is valuable not only for business insights and decision frameworks but also as a product. Within the context of an enterprise IoT solution, data analytics and artificial intelligence can be used to identify trends and anomalies in the underlying business process monitored/controlled by the IoT solution. This analysis can be used to improve workflow efficiencies, reduce operational costs, and improve equipment maintenance, as discussed earlier. This data can also be shared with customers either directly or through **Application Programming Interfaces** (**APIs**) to the enterprise IoT application. This improves the customer experience with the IoT solution and allows customers to further develop and innovate around the IoT data with their customers.

Summary

In this chapter, we have hopefully provided a good overview of how IoT enables an enterprise's digital transformation and the leading enterprise markets where IoT is making a significant impact. In our review, we have addressed many of the business needs solved by implementing an IoT solution, including operational cost savings and improved efficiencies. In the following chapters, we will go into more detail on the specific components of an enterprise IoT solution with a focus on cellular technologies.

2
Understanding IoT Devices and Architectures

With a good foundational understanding of IoT and how IoT enables enterprise digital transformations, this chapter will introduce you to end-to-end enterprise IoT solution architectures and devices. For common wireless IoT technologies, a high-level IoT architecture is shown in *Figure 2.1*. In this network architecture, IoT devices used in the various IoT markets and use cases reviewed in *Chapter 1, Transforming to an IoT Business*, have a brokered gateway connection to an **Internet Protocol** (**IP**) network, which connects to the internet and the enterprise IoT platform and application. Both the IoT platform and application elements could reside in an enterprise data center or more typically, in the domain of a **cloud service provider** (**CSP**), such as **Amazon Web Services** (**AWS**) or Microsoft Azure, where the IoT data is ingested and formatted for display on an enterprise IoT dashboard.

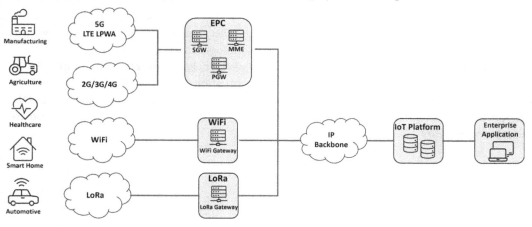

Figure 2.1 – High-level IoT architecture

In this chapter, we will provide an overview of the high-level enterprise IoT architecture shown in *Figure 2.1* with a focus on cellular IoT networks to include a description of the primary network components from the device to the IoT application. We will also provide an overview of the most common IoT device types and their associated use cases.

By the end of this chapter, you should have a good understanding of the high-level IoT architecture from the device to the enterprise cloud application including the main architecture components and protocols. Moreover, you should gain a good understanding of the IoT device types, their associated use cases, as well as some of the best practices for selecting both IoT devices and IoT solution architectures.

As such, we will cover the following topics:

- High-level IoT architecture
- Cellular IoT network architecture
- Overview of IoT device types and use cases
- Best practices

Let us begin our discussion with an overview of the high-level end-to-end IoT architecture shown in *Figure 2.1*, which applies to all IoT solutions. We will cover the cellular IoT architecture in detail later in the subsequent sections.

High-level IoT architecture

Regardless of the IoT use case, the basic architecture shown in *Figure 2.1* applies. At a high level, all IoT solutions have a device component for collecting the relevant IoT data which includes sensor data such as location, temperature, acceleration, flow, or other remote monitoring information. This data is typically transmitted using various wireless technologies, such as Wi-Fi or cellular, to an IP gateway using specific messaging protocols to the IoT platform and application across an IP-based network. Most IoT solutions also require two-way communication from the IoT application to change, for example, the device configuration or affect a change to an actuator on a machine. To explore the components of this high-level architecture in more detail, let us look at the five-layer IoT protocol stack shown in *Figure 2.2*.

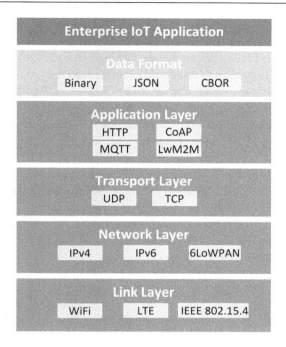

Figure 2.2 – The IoT protocol stack

This is a more detailed look at the high-level IoT architecture discussed in *Chapter 1, Transforming to an IoT Business*. This IoT protocol stack includes the basic technologies to move data from the IoT device to the enterprise IoT application.

Let us start our review with the link layer, which is the physically wired and/or wireless technology integrated into the IoT device that enables communication with the enterprise IoT application.

The link layer

This is the lowest layer in the IoT IP protocol stack and it includes various physical radio technologies such as **Wireless Fidelity (Wi-Fi)**, which is based on the IEEE 802.11 family of standards, **Long-Range Wide Area (LoRa)**, and **Low-rate Wireless Personal Networks (LR-WPANs)**, which are based on the IEEE 802.15.4 family of standards. These radio technologies, which are most commonly used in enterprise IoT solutions, will be covered in *Chapter 3, Introducing IoT Wireless Technologies*. This layer of the IoT stack also includes the various cellular technologies that are the focus of this book. We will review all of the LTE and 5G radio technologies in detail in *Chapter 4, Leveraging Cellular IoT Technologies*, and *Chapter 5, Validating 5G with IoT*. It is important to note this layer also includes wired Ethernet connections, which were standardized in 1983 as IEEE 802.3. These link layer technologies define how the device accesses the IP backbone network, which brings us to the network layer.

The network layer

The network layer defines how the data from the devices is packetized with a header and body and routed over the internet, primarily using standard IP either **version 4 (v4)**, **version 6 (v6)**, or **IPv6 over Low-Power Wireless Personal Area Network (6LoWPAN)**. The IP packet contains information about the content, source, and destination of the data packet for routing. Each IP packet has a unique address with IPv4 first deployed in 1982, providing around 4.3 billion unique IP addresses. Given the pool of unused IPv4 addresses is nearly exhausted, most networks are moving from IPv4 to IPv6, which provides about *3.4 X 10*^{38} addresses, which is essentially unlimited. The 6LoWPAN standard was defined by the **Internet Engineering Task Force (IETF)**, specifically for IEEE 802.15.4 (LR-WPAN) devices but is now being adapted for other low-power, constrained wireless technologies such as Bluetooth and Wi-Fi. 6LoWPAN allows interoperability with standard IPv6 and provides a method of converting the standard IPv6 packets into a format that can be handled by these constrained devices while also supporting security. This brings us to the transport layer, which provides the mechanism for sending the device message without errors.

The transport layer

There are two primary IP protocols to allow devices to exchange data with the IoT application. These are the **Transmission Control Protocol (TCP)** and the **User Datagram Protocol (UDP)**. TCP is a connection-oriented protocol that establishes and maintains a connection between the device and IoT application until the data exchange is completed. It manages data flow control between the application layer and lower layers. An important feature of TCP is that it includes error detection and acknowledgments that the data packets have arrived. As such, TCP can retransmit and reorder data packets accordingly, introducing network latency, which can impact latency-sensitive applications such as **Voice over IP (VoIP)**. On the other hand, UDP is a connectionless protocol with no way to detect if the data exchange is completed and does not include acknowledgments or error correction of failed packet deliveries. As such, UDP is a much simpler data exchange protocol with lower latency compared to TCP. TCP is used for most internet and IoT data traffic with application protocols such as **Hypertext Transfer Protocol (HTTP)**. UDP is appealing to enterprise IoT solutions because it requires fewer network resources and does not need to maintain a constant connection with the constrained IoT device which is low power with limited computing resources. This brings us to the application layer which is responsible for the formatting and presentation of the IoT data to the enterprise IoT application.

The application layer

The application layer is analogous to an internet browser that implements application layer protocols such as HTTP or **File Transfer Protocol (FTP)**. In the context of IoT solutions with constrained devices (low-power and limited computing power), HTTP requires too much device computing power which is the impetus for alternative IoT-specific protocols such as **Message Queuing Telemetry Transport (MQTT)**, **Constrained Application Protocol (CoAP)**, and **Lightweight Machine 2 Machine (LwM2M)**.

There are several other protocols that could be used in your enterprise IoT solution, but these are the most common for low-power, constrained IoT devices. Let us now review the main features of each of these protocols that make them good choices for enterprise IoT solutions.

MQTT

MQTT is a publish-subscribe communication protocol designed for sensor networks. It allows devices to send or publish data on a given topic to a server. There is an MQTT broker between the publisher and subscriber, which transfers data to the devices/servers that are subscribed. The IoT sensor devices can be thought of as publishers in the network, with the client IoT application being the subscriber to receive this data through the MQTT broker. This protocol is commonly used in enterprise IoT solutions and is supported by popular IoT platforms such as AWS and Microsoft Azure. It is easy to implement, and, as a publish-subscribe model, it is also delay/latency tolerant. The main disadvantages of a device using MQTT are the high power consumption, because MQTT uses TCP, and the lack of encryption. Due to this drawback, there is also a version of MQTT called **MQTT for Sensor Networks (MQTT-SN)** which is a lighter-weight protocol with reduced message payload size that uses UDP instead of TCP as the transport protocol. This version of MQTT is much better for battery-powered IoT devices with limited compute processing and storage.

CoAP

Unlike MQTT, CoAP uses a two-way client-server architecture such as HTTP with UDP instead of TCP. Like MQTT, it is a lightweight protocol designed for constrained, low-power IoT networks. CoAP has small message sizes and uses **Datagram Transport Layer Security (DTLS)** over UDP for security. Although it uses UDP, the client-server data exchange can also be set for transmission acknowledgment for more reliability. The CoAP sensor devices are the servers on the IoT network that are routable with a one-to-one connection with the client IoT application. The IoT application can get, delete, or post data to the CoAP sensor devices.

LwM2M

LwM2M is another simple, lightweight protocol for constrained IoT networks that can be used for IoT device management and telemetry. It was standardized by the **Open Mobile Alliance (OMA)** in 2017. It was originally built on CoAP as the underlying two-way client-server architecture, so it inherits the benefits of CoAP with DTLS over UDP for security. LwM2M was designed to support both IoT device management and telemetry with a defined hierarchical object and resource structure stored on the IoT device, as shown in *Figure 2.3*. The IoT device sensor data (location, temperature, etc.) can be captured at the resource level by the server device and read by the client IoT application issuing a read operation on the device parent object which could be *location* or *sensor readings*.

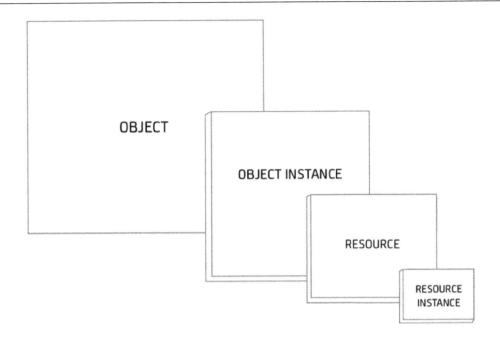

Figure 2.3 – The LwM2M object and resource structure

Using this object-resource structure, the protocol specification also supports many normal IoT device management functions, such as firmware and software updates and connectivity monitoring/management. LwM2M has the advantage of being a lightweight protocol that can support both device management and reporting.

With the various combinations of these application layer protocols with the transport, network, and link layers, the format of the IoT data payload is the last layer in our review of the high-level IoT architecture.

The data format

Due to the constrained computing power of most IoT devices, there are three main lightweight data formats used in IoT device messages to the IoT application server. These are binary, **JavaScript Object Notation (JSON)**, and **Concise Binary Object Representation (CBOR)**. The raw IoT data is normally formatted on the device prior to transmission but could be transformed/parsed on the enterprise IoT platform for ingestion in the IoT application. Although all data payloads are technically binary as they are all a collection of bytes, the difference between *binary* and *text* data formats is what those bytes represent. In *text* data formats such as JSON, the bytes represent standard text characters commonly known as **American Standard for Information Interchange** (**ASCII**), but in binary data formats, the bytes represent custom encoded and compressed data. A binary data format minimizes the payload size and adds to the security of the data due to the need for a custom decoder. JSON is one of the most common lightweight, text-based IoT data interchange formats that is based on the JavaScript programming language. It is easy for humans to read/write and easy for machines to parse and generate. As shown in *Figure 2.4*, the IoT data in JSON format is stored in key/value pairs separated by a comma. JSON values can be strings, numbers, boolean (written as `true` or `false`), objects (wrapped in curly braces), or arrays (wrapped in brackets) of any type.

```
{

"Device" : "Temp Sensor",

"Battery_Power" : "Good",

"Firmware_Version" : "123",

"Sensor_Data" : {

"Location" : "Somewhere, USA",

"Temperature" : "30 deg Celsius"

}

}
```

Figure 2.4 – JSON data format example

CBOR is a binary data format loosely based on JSON that uses the same key-value pairs as JSON but in a more concise binary format, which increases processing speed on the IoT device and minimizes the data payload size.

With a high-level overview of the high-level IoT architecture protocol stack, let's review the cellular IoT network architecture components. We will review LTE technology in *Chapter 4, Leveraging Cellular IoT Technologies*, and the newest cellular 5G technology in *Chapter 5, Validating 5G with IoT*.

The cellular IoT network architecture

A high-level diagram of the LTE network architecture in the context of IoT solutions is shown in *Figure 2.5*. Starting with the IoT device, we will discuss the process for accessing and sending data on the LTE network and the role of the network components shown.

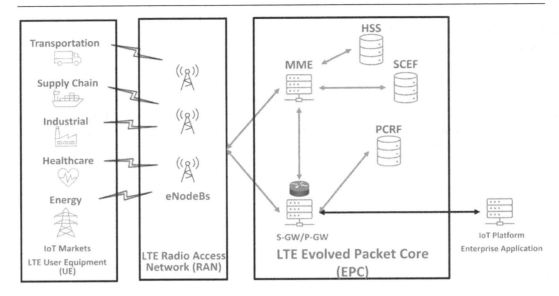

Figure 2.5 – The cellular IoT network architecture

In this section, we will describe each of the main cellular network components shown in *Figure 2.5*, including the role of each network element in identifying and allowing cellular IoT devices to access the network.

Let us begin our overview with the IoT device that is used in the enterprise IoT use case, which is termed **user equipment** (**UE**).

The UE

The UE or IoT device must be compatible with the LTE network to gain access. This means it must support the specific LTE technology and frequency bands and have the properly licensed LTE carrier credentials (for example, AT&T or Verizon), which are stored in the device's **Subscriber Identity Module** (**SIM**). The UE compatibility requirements will be discussed in more detail in *Chapter 4, Leveraging Cellular IoT Technologies*. In an IoT network deployment, the compatible UEs will connect with the closest LTE **Evolved Node B** (**eNodeB**) base station where there is radio coverage and request network services including data, **voice** (**VoLTE**), and/or **Short Message Service** (**SMS**), also known as text messaging. Data service is the most common service required for IoT devices. VoLTE service is an IP-based technology enabled by the **Evolved Packet Core** (**EPC**) shown in *Figure 2.5* and the **IP Multimedia Subsystem** (**IMS**), which is a separate standalone network element connected to the EPC and is not shown in *Figure 2.5*. As such, IoT devices requiring voice on LTE need to support an IMS client application on the LTE module. SMS, which is used in many IoT solutions, could also be delivered over IMS, but some carriers deliver SMS over an LTE control channel through the **Signaling Gateway** (**SGW**), which is called SG-SMS. If SMS is required in your IoT solution, it is important to validate the SMS call flow with your selected network carrier. To support the delivery of small amounts of infrequent IoT data, some LTE network operators have implemented **Non-IP Data Delivery (NIDD)**, which is capable of sending 1500 bytes in a single transmission without the overhead of IP or TCP/UDP protocols. All of these network services may not be available for all LTE technologies such as LTE-M and NB-IoT, so it is important to validate the network services available from your selected network carrier for your IoT solution.

The UE or IoT device connects to the selected network carrier **Radio Access Network (RAN)** via the LTE eNodeB or base stations, which we will cover next.

eNodeB

The LTE eNodeBs or base stations provide the wireless radio coverage known as the RAN in the licensed LTE bands of various network carriers globally. This is the same radio network used by your smartphone and, like your smartphone, the UE will activate a **Packet Data Protocol (PDP)** context session with the network and wait for a response. As their core business, network carriers invest heavily in the deployment, maintenance, and support of their RAN to ensure the best coverage and performance for all devices on the network. The eNodeB forwards this UE request to the LTE **Evolved Packet Core (EPC)**, which is the *brain* of the LTE network.

EPC

The EPC validates the session request from the UE, generates a PDP context, and ultimately provides access to the **Packet Data Network (PDN)** for a connection to the IoT application server. To provide this access, the EPC uses the five following primary components:

- The **Mobility Management Entity (MME)**
- The **Home Subscriber Server (HSS)**
- The **Serving Gateway and PDN Gateway (S-GW/P-GW)**
- The **Policy and Charging Rules Function (PCRF)**
- The **Service Capability Exposure Function (SCEF)**

Let us review the roles of these components, starting with the MME.

The MME

The MME is responsible for managing the UEs on the network, including session management, subscriber authentications, and roaming/handovers between networks. The MME authenticates the UE with the user's home HSS.

The HSS

The HSS is the master user subscription database, which includes the user's **international mobile subscriber identity (IMSI)** and mobile telephone number. It also generates the network security information from the UE identity keys that are provided to other network entities. The HSS also has information about the PDN to which the device is trying to connect and has the **Access Point Name** (**APN**) or identifier to the internet associated with the UE, which defines the allowed PDN connections. This APN could be either a public or private APN. A public APN is used by many different subscribers/customers on a carrier network, whereas a private APN is private to one customer and can be customized for a highly secure **Virtual Private Network** (**VPN**) connection or tunnel to an enterprise network. A private APN can also be customized to provide either static or dynamic IP addresses to the UE. A private APN is the preferred access type for IoT devices because of the inherent privacy and security it provides.

The S-GW/P-GW

The S-GW receives instructions for the MME to set up or tear down data sessions for UEs and acts as the interface between the P-GW and MME. It also manages the IP packets between the P-GW and eNodeB. The P-GW provides access to external PDNs such as the IoT application server network and acts as an IP router with router tunneling and signaling protocols.

The PCRF

The PCRF server interfaces with the P-GW and manages the carrier service policy for the network and individual UE subscribers, linking carrier charging rules with the subscriber data/bandwidth usage, network prioritization, and **Quality of Service** (**QoS**). The PCRF allows network carriers to differentiate data services while optimizing revenue.

The SCEF

The SCEF network element provides an interface to the IoT application server to expose network interfaces enabling carrier-specific IoT functions, such as NIDD for low-power devices and IoT device triggering when the IP address is not reachable. These are carrier-specific functions not implemented by all LTE carriers.

We will cover the LTE network architecture in more detail in *Chapter 4*, *Leveraging Cellular IoT Technologies*. With a high-level understanding of the IoT architecture, protocol stack, and cellular network architecture, we will now provide an overview of IoT device types and the most common associated use cases.

An overview of IoT device types and use cases

To support the enterprise IoT markets described in *Chapter 1*, *Transforming to an IoT Business*, there are various IoT devices that could be selected depending on the specific use case. In general, IoT devices can be categorized into the following five categories, which cover most enterprise IoT use cases:

- Serial modems
- Routers

- Gateways

- Remote monitoring devices

- Asset trackers

We will go into more detail on cellular (LTE) IoT devices in *Chapter 6, Reviewing Cellular IoT Devices with Use Cases*.

Serial modems

A serial modem is the most basic and common IoT device supporting many enterprise IoT use cases, where the interface to the monitored/controlled machine/appliance is one of the following standardized serial protocols:

- **Universal Serial Bus (USB)**

- **Recommended Standard 232 (RS-232)**

- **Recommended Standard 485 (RS-485)**

- **Serial Peripheral Interface (SPI)**

- **Inter Integrated Circuit (I2C)**

In the context of an enterprise IoT solution, a serial modem enables two-way IP wireless communication with endpoints such as factory equipment, transportation systems, **Supervisory Control and Data Acquisition (SCADA)** utility infrastructure, building automation systems, lighting controls, and agriculture control systems. Through this two-way communication, an enterprise can not only collect critical data on the endpoints, but they can also affect a change in the endpoint parameters based on that data. For example, in monitoring a machine in a factory, data indicating maintenance is due may cause the operator to shut down the machine, schedule the maintenance, and change the production workflow accordingly. As another example, refrigerators/coolers in a retail store can typically be monitored for temperature variations that are out-of-range via a serial RS-232/RS-485 modem connected to the individual coolers, or the cooler management system installed in the store. In the context of IoT solutions with SCADA, which has been termed the *Smart Grid*, serial modems are commonly used for remote monitoring of electric and gas meters (*Smart Metering*), powerlines, and transformers which improves the operational efficiency of utilities.

Routers

A router is a device with at least one ethernet port for routing IP traffic over a **Wireless Local Area Network (WLAN)**, **Wireless Wide Area Network (WWAN)**, or both. Routers typically support both cellular LTE/5G WWAN and WLAN technologies and provide a direct two-way IP connection to an enterprise endpoint, without the need for any special device drivers or protocols. Unlike serial modems, routers normally support higher data throughputs and have built-in security features (e.g. VPN, built-in firewall, and traffic management) to protect/manage the transported data. Routers are commonly used with enterprise endpoints, such as laptop and desktop computers as well as ethernet LAN hubs at enterprise branch offices, to provide wireless failover for wired WANs or even the primary WAN edge. In many enterprise IoT solutions, routers serve the function of directly backhauling IP data to the cloud and backend enterprise platforms. For example, in point-of-sale portable kiosks or vehicles where wired internet connectivity is not possible, routers commonly provide both the WLAN and WWAN backhaul connectivity.

Gateways

Gateways are IoT devices that translate an IoT WLAN technology such as Bluetooth, Wi-Fi, or LoRaWAN to a WWAN technology such as ethernet or cellular LTE/5G. We will cover the IoT WLAN technologies in more detail in *Chapter 3, Introducing IoT Wireless Technologies*. In practice, gateways convert IoT wireless sensors and remote monitoring data into IP traffic sent over WWAN for many enterprise IoT solutions. Like a router, gateways backhaul data over WWAN technology, but unlike routers, gateways integrate a WLAN sensor technology instead of wired ethernet. Gateways are commonly used in enterprise IoT applications where the data from several wireless sensors (e.g. temperature, humidity, vibration, light, etc.) are aggregated together and sent via WWAN to the enterprise cloud application. Gateways with improved computing power can also process data locally, which has been termed *edge compute* to reduce the WWAN traffic and make near real-time decisions on the data locally. We will cover edge-computing and machine learning in more detail in *Chapter 6, Reviewing Cellular IoT Devices with Use Cases*.

Remote monitoring devices

The hallmark of IoT has always been the billions of connected things where these "things" are mostly wireless sensors, including sensors for temperature, humidity, vibration, air quality, light, and motion. Remote monitoring devices are devices that integrate wired or wireless sensors to monitor machines or environments for anomalous behavior. Unlike gateways, remote monitoring devices are typically standalone WWAN (e.g. LTE/5G, LoRaWAN) devices with integrated sensors. They are normally very simple, low-cost devices, which have limited compute processing capabilities and deliver time-series IP data at specified intervals directly to the enterprise cloud application for processing. For example, air quality sensors deployed throughout a city can provide real-time air-quality assessments and warnings. In the context of smart cities, another good use case for remote monitoring devices is structure monitoring where remote monitoring of structures such as bridges, buildings, stadiums, and ships for vibrations and issues with their structural integrity can be used to prepare timely corrective action plans. In the agriculture market, soil sensors deployed as part of an agriculture enterprise IoT solution can optimize the efficiency of irrigation and fertilizer operations saving both water and money. With the advent of **low-power wide area** (**LPWA**) technologies, remote monitoring IoT devices are becoming more popular, as they remove the need for an intermediate WWAN gateway, which further reduces the cost of an enterprise IoT solution.

Asset trackers

Asset-tracking IoT devices primarily report endpoint location and condition information and are commonly used in the transportation and supply chain markets. There are four methods commonly used for location resolution:

- **Global navigation satellite systems (GNSS)**
- Wi-Fi access points
- Bluetooth beacons
- Cellular information

GNSS is a general term for any satellite constellation that provides positioning, navigation, and timing services. While the US-owned **Global Positioning System (GPS)** and Russian-owned **Global Navigation Satellite System (GLONASS)** are the most widely used satellite global geolocation systems, there are also other navigation systems, such as **BeiDou Navigation Satellite System (BDS)**, which is owned and operated by China, and Galileo, which is owned and operated by the European Union. All of these geolocation systems provide geolocation and time information to a GNSS receiver for that specific system anywhere on the Earth where there is an unobstructed view of four or more satellites. Asset tracking devices with an integrated GNSS radio can provide location accuracy within four meters with an unobstructed view of the sky. Since the asset being tracked is typically inside a building or vehicle, it is common for asset trackers to include other location technologies.

For indoor locations, the Wi-Fi **access points (APs)** in the vicinity of an asset tracker with an integrated Wi-Fi receiver can be used to resolve a location within 20 meters. Asset tracking devices with Wi-Fi can report the unique **media access control (MAC)** addresses and **radio frequency (RF) received signal strength indicator (RSSI)** of the Wi-Fi APs, which are used by a cloud-based, crowd-sourced location service application such as Google or Azure to provide a rough latitude/longitude location of the asset. The accuracy of this location is not as accurate as GNSS but is usually good enough to determine if the asset is in a specific building.

In some enterprise IoT solutions, a more accurate location of equipment and products in a warehouse or distribution center is required. In this case, Bluetooth can be used to provide location accuracies of less than five meters, which approaches GNSS accuracy. This location solution requires the device to have an integrated Bluetooth receiver in conjunction with several Bluetooth location beacons deployed inside the building. Like the Wi-Fi location solution, the asset tracker will report the MAC addresses of the Bluetooth beacons and signal strength to a cloud-based application that resolves the location in the building. The main disadvantage of this location solution is the cost of deploying the Bluetooth beacons in the building.

In cases where GNSS, Wi-Fi, and Bluetooth are not available, cellular information based on nearby cell towers can also be used to provide a rough location for the asset being tracked. In this case, the cellular-based asset tracker reports the signal strength of the various cell towers it sees to a cloud-based application that uses an algorithm to resolve the location within about 500 meters. This accuracy is generally good enough to determine that an asset is in a certain building. The disadvantage of this location method is the low accuracy, but the advantage is a lower device cost where the device does not need a GNSS or Wi-Fi receiver.

Many of the enterprise IoT solutions that rely on asset trackers use multiple location methods to provide a more robust solution in many environments. For example, assets shipped in large metal shipping containers globally encounter many different RF environments along the supply chain and may not have any RF signal at various stages of their journey, so implementing multiple location solutions provides better tracking visibility.

To conclude our discussion of IoT architectures and devices, we will now review some of the best practices when choosing IoT devices and architecture technologies for your enterprise IoT solution.

Best practices

In designing your enterprise IoT solution around a particular IoT use case, it is important to first define the strategic business model or problem solved with the solution that drives business value. Along the same lines, it is important to define the **key performance indicator** (**KPI**) metrics for how the solution is evaluated. For example, is your IoT solution solving a business problem such as inventory/warehouse management, predictive maintenance in a smart factory, or fleet management? Alternatively, is your IoT solution creating a new business model with subscription services to your customers such as asset tracking, remote healthcare, security, remote monitoring, or home automation? With a clear scope for your enterprise IoT solution, it is important to then identify the technologies from the IoT device and the WWAN to the IoT protocol stack and architecture. In this regard, here are five best practices for developing your enterprise IoT solution:

- Leveraging *off-the-shelf* technology components
- Leveraging established IoT technology partners
- Planning on solution life cycle management
- Slow and steady wins the race
- Security is paramount

Leveraging off-the-shelf technology components

In general, it is a best practice to not try and reinvent the wheel but select off-the-shelf components with standardized technologies that allow for better support and flexibility. For example, in the case of asset tracking/monitoring, there are a myriad of asset-tracking devices with various sensors in many different configurations depending on the asset that is tracked, and there are many commercial cloud-based asset-tracking platforms and applications. Given this, it would not necessarily make sense to develop your own enterprise-asset-tracking platform, protocol stack technology, or asset-tracking device. Instead, it makes more sense to reuse and customize the available commercial technologies where these are already available. In other words, it is good to identify the unique features of your enterprise IoT solution not being met in commercial solutions and focus on filling these gaps where there is business value.

Leveraging established IoT technology partners

Another best practice is to work with a few IoT partners that can support the more critical and complex technology components of your IoT solution such as WWAN connectivity and the IoT cloud platform/application. It rarely makes sense for your business to build and deploy your own WWAN network or develop a new IoT platform, as opposed to working with a licensed network operator such as AT&T and a cloud service provider such as Microsoft Azure. In this case, the IoT device, protocol stack, and application can be customized to meet the requirements of your enterprise IoT solution. For the critical components in your IoT solution, it is important to leverage established technology partners that can support the full life cycle of your solution.

Planning on solution life cycle management

We will cover the enterprise IoT solution life cycle management in detail in *Chapter 9, Managing the Cellular IoT Solution life cycle*. Most enterprise IoT solutions have a life cycle of well over five years, which means the IoT solution needs to have a mechanism for life cycle management with device management for firmware/configuration updates, device behavior monitoring, health reporting, performance metrics, and WWAN network health. Especially with regards to the WWAN and cloud-based IoT platform/application support, there should be **Service Level Agreements** (**SLAs**) where possible as well as full life cycle support built into your enterprise IoT solution. Since technological changes will occur, it is important that your enterprise solution is future-proofed as much as possible to avoid any surprises or unexpected expenses.

Slow and steady wins the race

When deploying your enterprise IoT solution, it is best to start with a **proof of concept** (**POC**) and a small-scale pilot to methodically evaluate with your team if the solution meets your business goals. This will inevitably reveal gaps in the IoT solution, typically requiring iterations to fix in follow-on pilot field deployments. By skipping this process and doing a faster large-scale deployment of a new IoT solution, any small gaps will be magnified many-fold, incurring higher financial risk and potentially jeopardizing an otherwise good enterprise solution.

Security is paramount

As IoT deployments continue to grow to billions of connected IoT devices, the security of these devices and the data they send should be a paramount concern in any enterprise IoT solution. Given that IoT devices, especially low-cost, constrained devices, are typically not as secure from malware and hacking exploits as high-end consumer devices such as smartphones, it is a best practice to ensure the security of not only the device but also the end-to-end IoT network solution. This means there should be a multi-layered approach to security and data privacy from the device endpoint through the network and application layers. We will cover privacy and security in detail in *Chapter 7, Securing the Internet of Things*.

Let us summarize what we covered in this chapter and how this serves as the foundation for the remainder of the book.

Summary

We started our discussion by reviewing the high-level IoT architecture and the embedded IoT protocol stack technologies, leading us to a review of the LTE network architecture that supports IoT devices. We then covered an overview of the common IoT devices with the associated use cases and concluded with a discussion of best practices when deploying your enterprise IoT solution. Overall, this chapter should provide the reader with a good understanding of the high-level IoT protocol stack, the LTE network architecture, and common IoT device types.

In the following chapters, we will introduce the most common IoT WLAN and WWAN technologies and provide a more detailed description of LTE and LTE LPWA technologies in the context of enterprise IoT solutions. Beginning with *Chapter 5, Validating 5G with IoT*, we will explore how to specifically leverage the latest 5G technology in the context of IoT and provide more insight into LTE/5G IoT solution devices, architectures, and security. Starting from *Chapter 8, Implementing an IoT Solutions with Case Studies*, we will conclude our review of cellular IoT enterprise solutions with a few case studies and some best practices for managing your enterprise IoT solution life cycle.

3

Introducing IoT Wireless Technologies

In this chapter, we will describe the various IoT wireless technologies, including the primary licensed and unlicensed **Wireless Wide Area Network (WWAN)** and **Wireless Local Area Network (WLAN)** technologies used in enterprise IoT solutions. We will compare the technologies from a technical and cost perspective, as well as listing the pros and cons of each technology to provide a basis for selecting the best technology for specific enterprise IoT solutions. We will conclude with a discussion of best practices in implementing IoT wireless technologies in your IoT solution.

We will cover the following main topics in this chapter:

- Licensed versus unlicensed technologies

- Network topologies

- WWAN and WLAN IoT technologies

- Technical and cost comparisons

- Wireless technology use cases

- Best practices in deploying IoT wireless technologies

Licensed versus unlicensed technologies

To begin our exploration of IoT wireless technologies, it is important to understand the difference between licensed versus unlicensed wireless spectrum. Unlicensed spectrum is public wireless spectrum that can be used by organizations without getting permission from the **Federal Communications Commission (FCC)** in the US or the governing wireless regulatory body in the deployment region globally. The **International Telecommunication Union (ITU)** designated **Industrial, Scientific, and Medical (ISM)** unlicensed frequency bands for many uses beyond telecommunication. Examples of IoT technologies that use unlicensed spectrum or ISM bands include Wi-Fi, Bluetooth, **Long Range (LoRa)**, and Thread, as well as other proprietary technologies. We will go into detail on each of these technologies later in this chapter. From a **radio frequency** (**RF**) perspective, the lower the frequency the better the range/coverage—so, for example, technologies utilizing the 900 MHz band will have better coverage/range than the same technologies using the 2.4 GHz or 5.8 GHz bands. The availability of unlicensed ISM spectrum may vary by global region, so the available unlicensed ISM bands may not overlap in all global regions, which is an important consideration for enterprise IoT solutions deployed globally. In general, the unlicensed 2.4 GHz band is consistent across all global regions, which makes it a popular choice for Wi-Fi and Bluetooth technologies. The most common unlicensed spectrum bands used in IoT solutions are as follows:

- **900 MHz**: This frequency band provides the best coverage and is used by LoRaWAN and proprietary wireless devices/solutions in the utility and manufacturing markets. In the US, this band comprises 5 MHz of bandwidth between 896 and 901 MHz and another 5 MHz of spectrum between 935 and 940 MHz. As an example use case, utilities use the unlicensed 900 MHz band for their **Advanced Meter Infrastructure** (**AMI**) supporting electric, gas, and water metering applications in the US.

- **2.4 GHz**: Given the global availability of this unlicensed spectrum, this frequency band is used globally in Wi-Fi and Bluetooth devices, typically in indoor, short-range WLAN applications. This unlicensed ISM band comprises 100 MHz of spectrum between 2.4 GHz and 2.5 GHz. Nearly all IoT solutions using Wi-Fi and/or Bluetooth technologies use this unlicensed band globally.

- **5.8 GHz**: As with the 2.4 GHz band, this frequency band is used globally in Wi-Fi devices, typically in indoor and outdoor, short-range WLAN applications. This unlicensed ISM band comprises 150 MHz of spectrum between 5.725 GHz and 5.825 GHz. The extra 50 MHz of spectrum compared to the 2.4 GHz band enables applications with higher data throughput. Many IoT solutions using Wi-Fi use this unlicensed band globally.

While the vast majority of IoT solutions use unlicensed spectrum for connectivity, the lack of exclusivity with public wireless spectrum means there is a high potential for mutual interference and traffic congestion with other users, which could negatively impact the connectivity in an IoT solution. Another concern is security, as there is no consistent, inherent network security. The main advantage of using unlicensed spectrum for IoT connectivity is the fact that the spectrum is free to use if you don't consider the capital and operational cost of the wireless network infrastructure.

Licensed wireless networks offer the benefits of greater reliability and performance given that the spectrum is licensed to specific licensees such as cellular carriers, which manage the performance and quality of service on the network as part of their core business. As such, the users of these networks have a greater assurance of good wireless performance with little or no interference from unauthorized users. Moreover, the coverage, range, and capacity of commercial cellular networks are much higher than unlicensed technologies and are synonymous with WWAN. Cellular wireless carriers such as AT&T, Verizon, and T-Mobile have the bulk of licensed spectrum in the United States where they deploy their **long-term evolution** (**LTE**) and 5G WWAN services supporting their mobility and IoT customers. We will cover the licensed LTE and 5G cellular frequency bands and the associated LTE/5G technologies in *Chapter 4, Leveraging Cellular IoT Technologies*, and *Chapter 5, Validating 5G with*

IoT. As part of this service, cellular carriers take on the ongoing operational cost of maintaining and upgrading the network to provide the best experience to their customers. The device data subscription costs are typically based on monthly data consumption, which is usually low for most IoT solutions.

Another important consideration when choosing licensed versus unlicensed technologies for your enterprise IoT solution is the network topology.

Network topologies

As shown in *Figure 3.1*, there are 3 primary network topologies used in wireless networks, with the mesh and star topologies being the most common:

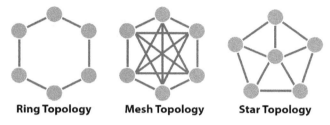

Figure 3.1 – Primary IoT wireless network topologies

Each IoT WLAN and WWAN technology uses specific network topologies, and there are advantages and disadvantages to each topology. For example, mesh topologies used with many **Wireless Personal Area Networks** (**WPANs**) and recent WLAN technology releases have the main advantage of network resiliency where the failure of one device does not disrupt connectivity due to redundant connections. The disadvantages of a mesh topology are the complexity of the network and device setup, as well as latency issues in traversing the network. In the case of star topologies used in cellular and WLAN technologies, the main advantages are reliability, efficiency, and centralized network management. The primary disadvantages of star topologies are the cost, complexity, and maintenance of the central nodes. We will refer to these network topologies in the context of the IoT WLAN and WWAN technologies later in this chapter.

Overall, there are applications for using both licensed and unlicensed technologies in enterprise IoT solutions. We will cover the technical and cost trade-offs between the licensed and unlicensed WWAN and WLAN technologies later in this chapter, but first, let's explore the common IoT WLAN and WWAN technologies and how these technologies can complement each other in an enterprise IoT solution.

WWAN and WLAN IoT technologies

WWAN technologies include cellular LTE and 5G, as well as some unlicensed **Low Power Wide Area** (**LPWA**) technologies such as LoRaWAN, whereas WLAN includes the prevalent Wi-Fi and Bluetooth technologies. Well over 80% of the connectivity in current IoT solutions are the WLAN and WPAN technologies, which are critical parts of the IoT ecosystem and will be reviewed next. We will cover licensed cellular LPWA in detail in *Chapter 4*, *Leveraging Cellular IoT Technologies*.

There are four primary WLAN/WPAN unlicensed technologies used in most enterprise IoT solutions. These are as follows:

- Bluetooth/**Bluetooth Low Energy (BLE)**
- Wi-Fi
- LoRa and LoRaWAN
- LR-WPAN (IEEE *802.15.4*)

We will start our review of WLAN/WPAN technologies with Bluetooth/BLE, which is the most prevalent WPAN technology.

Bluetooth/BLE

Bluetooth is a ubiquitous, short-range, standardized WPAN technology developed by the **Bluetooth Special Interest Group** (**Bluetooth SIG**) and introduced in 1998. It operates in the 2.4 GHz unlicensed frequency band and is complementary to Wi-Fi, as we will discuss in the next section on Wi-Fi. This technology is integrated into almost every smart consumer device such as phones and tablets and is normally used for wireless speakers/headphones and smart home applications.

Typically, Bluetooth supports data rates of up to 2 Mbps and a range of around 10 meters, but Bluetooth has evolved over the past 24 years to BLE, introduced in Bluetooth 4.0, and **Bluetooth Mesh**, introduced in Bluetooth 5.0, which can support mesh instead of star topologies. BLE supports lower power and longer ranges of around 240 meters compared to classic Bluetooth. BLE was introduced in Bluetooth 4.0 in 2010 and typically uses 1 to 5% of the power of classic Bluetooth. This allows for BLE sensor devices that can be powered by coin-cell batteries or harvested energy (for example, thermoelectric sources). We will discuss the common Bluetooth/BLE IoT use cases later in this chapter. Overall, Bluetooth and BLE will continue to be key WPAN/WLAN wireless IoT technologies, supporting many remote monitoring and asset-tracking enterprise IoT use cases.

Wi-Fi

Wi-Fi is another ubiquitous, short-range, standardized WLAN technology developed by the Wi-Fi Alliance and introduced in 1997. It operates in both the 2.4 GHz and 5.8 GHz unlicensed frequency bands. Wi-Fi is based on the IEEE *802.11* family of standards. As with Bluetooth, Wi-Fi is integrated into nearly all consumer devices and many IoT routers and gateway devices for "hotspot" WLAN coverage.

Wi-Fi technology has evolved over the past 25 years from supporting 1-2 Mbps to 40 Gbps with the pending release of Wi-Fi 7. The evolution of the Wi-Fi IEEE *802.11* standards through Wi-Fi 6 is shown in the following table:

Standard	Meaning	Year
802.11	Standard WiFi	1997
802.11 b		1999
802.11 a		1999
802.11 g	WiFi 3	2003
802.11 n	WiFi 4	2009
802.11 ac	WiFi 5	2014
802.11 ad	Milimeter wave	2010
802.11 ah	WiFi Halow	2017
802.11 ax	WiFi 6	2019

Figure 3.2 – Wi-Fi standard evolution

Generally, Wi-Fi has a longer range than Bluetooth, around 45 meters indoors and 100 meters outdoors. Compared to Bluetooth, Wi-Fi can use as much as 40 times the power of Bluetooth, which is an important consideration for battery-powered devices used in an enterprise IoT solution. Moreover, the cost of Wi-Fi chipsets/modules is significantly higher than Bluetooth, which is an important factor in low-cost sensor solutions.

As with BLE and Bluetooth Mesh in the Bluetooth standards, Wi-Fi standardization has evolved to meet the needs of the IoT market. For example, IEEE *802.11ah*, which has been branded **HaLow**, was ratified in 2016 and allows for lower power and longer ranges of up to 1,000 meters (at the cost of lower data throughput) competing directly with LPWA technologies such as LoRa and cellular LTE-M. HaLow uses the 900 MHz unlicensed spectrum for an extended range. HaLow has the advantages of being a standardized Wi-Fi technology with a robust ecosystem of vendors and offering the features of other LPWA technologies.

With its prevalence and broad portfolio of standardized profiles, Wi-Fi will continue to play a critical role in enterprise IoT solutions. Wi-Fi can support both broadband and narrowband IoT applications with varying levels of power consumption using a common, standardized platform that is ubiquitous. Especially with the deployment of low-latency, high-bandwidth 5G cellular technologies, the more recent high-bandwidth Wi-Fi standard releases are the perfect complementary WLAN technology.

LoRa and LoRaWAN

LoRa is a proprietary WLAN and LPWA technology introduced in 2015 that operates in the unlicensed 900 MHz frequency band (902-928 MHz) band in North America and similar sub-GHz bands globally. As with Wi-Fi, Bluetooth, and cellular, LoRa uses a star network topology. LoRaWAN uses the LoRa physical radio layer and refers to the networking protocol for WAN coverage. It is a proprietary technology in that the LoRa and LoRaWAN intellectual property as well as the development of the radio chipsets are owned and developed by Semtech, which is a semiconductor company based in the US. The ongoing development of the LoRaWAN protocol is managed by the non-profit LoRa Alliance where Semtech is a founding member.

LoRa and LoRaWAN technologies provide low-power, low data-rate, bi-directional communication with IoT devices globally and are primarily used with low data-rate sensors and remote monitoring devices. LoRa can support data rates of between 0.3 Kbps and 50 Kbps, which is significantly lower than either Bluetooth or Wi-Fi. The power consumption of LoRa is much lower than Wi-Fi and, as with BLE, makes it a good choice for enterprise IoT solutions with low-power sensor endpoints. LoRa can support ranges of around 10 km, which means it could be deployed as either a WLAN technology with WWAN (cellular LTE or 5G) gateway backhaul or as a WWAN LPWA technology such as cellular. When deployed as an LPWA technology, LoRaWAN can offer excellent coverage but with very low data rates compared to cellular LPWA technologies. We will cover the technical comparisons of LoRa with other WLAN and WWAN technologies later in this chapter.

Overall, LoRa provides a good, low-power, long-range technology enabling many enterprise IoT solutions with low data-rate requirements, but it requires the deployment and support of the proprietary LoRa network infrastructure to enable an LPWA IoT solution. As a low-power, long-range technology, LoRa will continue to play a critical role in enterprise IoT solutions such as **industrial IoT (IIoT)** and remote sensor monitoring applications.

LR-WPAN (IEEE 802.15.4)

The **Low-Rate WPAN (LR-WPAN)** technology standard was defined by the IEEE *802.15* working group in 2003. As shown in the following standard protocol stack diagram, LR-WPAN is the base radio protocol for the ZigBee, WirelessHART, 6LoWPAN, and Thread specifications. Where ZigBee and WirelessHART specify the network, transport, and application layers, Thread only specifies the network, transport, and application layers:

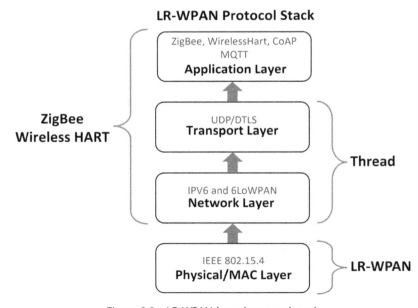

Figure 3.3 – LR-WPAN-based protocol stack

All the LR-WPAN technologies operate primarily in the unlicensed 900 MHz and 2.4 GHz frequency bands and support data rates of around 250 Kbps with a typical range of 10 meters, so they are well suited for low data-rate, enterprise IoT solutions. These technologies also use a mesh network topology for self-healing network resiliency. The power consumption and cost of the endpoint devices (typically sensors) are like BLE and LoRa devices, with special low-power/sleep modes included in the technical specifications. All these technologies rely on IoT gateway devices for the WWAN backhaul of data to the enterprise IoT cloud application.

Enterprise IoT solutions based on ZigBee, WirelessHART, and Thread have been commonly deployed in both industrial and home automation applications. With the promise of enabling the connection of billions of IoT devices, these standardized technologies have strong industry ecosystem support and are enabling use cases such as factory machine monitoring, industrial field operations, and smart home automation/monitoring. As such, these technologies should be considered in your enterprise IoT solution, especially with IIoT and smart home applications.

Technical and cost comparisons

There is not a single WLAN/WPAN or WWAN technology that will fit all enterprise IoT solutions, as there are technical and cost trade-offs with each technology. In this section, we will explore these trade-offs between the various IoT wireless technology choices. In *Figure 3.4*, we compare the WLAN/WPAN technologies discussed earlier with cellular LPWA where we chose LTE-M as the comparison cellular technology:

Feature	Bluetooth/BLE	Wi-Fi	LoRa	LR-WPAN	Cellular LPWA (LTE-M)
Frequency Spectrum	Unlicensed	Unlicensed	Unlicensed	Unlicensed	Licensed
Range	10 m – 240 m*	10 m – 1000 m**	2 km – 20 km	10 m – 100 m	1 km – 10 km
Power Consumption	Very Low	Medium	Low	Low	Medium
Data Throughput	125 kbps – 2 Mbps	1 Mbps – 1.3 Gbps	10 kbps – 50 kbps	20 kbps – 250 kbps	Up to 1 Mbps
Module Cost	Very Low	Low	Low	Low	Low
Network Topology	Star/Mesh	Star	Star	Mesh	Star
Data Category	Sensor Data	Data, Voice, HD Video	Sensor Data	Sensor Data	Sensor Data, Voice

*With Bluetooth 4 long range
**With IEEE 802.11ah (HaLow)

Figure 3.4 – IoT wireless technology comparison

As shown in *Figure 3.4*, the range, power consumption, and data throughput vary across technologies. There is an inherent trade-off between these features where, for example, it is possible to increase range by lowering data throughput and vice versa. Given these trade-offs, it is important to determine the required coverage/range, power budget, and data throughput for your enterprise IoT solution to select the optimal wireless technology. Let's explore the frequency spectrum, range, power consumption, data throughput, and module cost of each of these wireless technology features in more detail.

Frequency spectrum

As discussed earlier in this chapter, there are considerations in choosing unlicensed versus licensed technologies. There are essentially three areas to consider in choosing unlicensed spectrum technologies, which are as follows:

- Interference
- Infrastructure cost and availability
- Network capacity

The most important consideration in choosing unlicensed spectrum technologies is interference with other users of the spectrum.

Interference

With any unlicensed spectrum technology, there is the possibility of interference from other users, which can impact the performance (latency and data throughput) of the wireless connection in your IoT solution. With the increase in IoT solutions using unlicensed spectrum, this interference will be an ongoing problem in the future. There are techniques to manage this interference such as frequency-hopping, where the radio transmitter automatically selects a specific block of spectrum or channel that is relatively interference-free, but this should be a risk factored into your technology selection and IoT solution implementation. On the other hand, licensed spectrum is managed and maintained by the spectrum owners, such as the cellular network operators, as part of their service offering to their customers. There is a cost for this management, but it is typically shared with all customers and included in the customer-contracted service offerings.

Infrastructure cost and availability

Although unlicensed spectrum is free to use, there is the issue of who owns, deploys, and maintains the unlicensed spectrum network infrastructure, and there is a capital and operational cost for this infrastructure that needs to be factored into the enterprise IoT solution planning. This cost will need to be amortized over the life cycle of the IoT solution, discussed in *Chapter 9, Managing the Cellular IoT Solution Life Cycle*. Moreover, the availability of the unlicensed network infrastructure needs to be guaranteed over the full IoT solution life cycle, which could be 10 years or more. Both the unlicensed network infrastructure capital/operational cost and availability are critical factors often underestimated in enterprise IoT solution planning. Using licensed cellular service inherently includes the infrastructure costs and availability in the customer-contracted service offerings.

Network capacity

Licensed cellular is a service where the carrier manages all aspects of the service, including the number of users that can be supported, also known as **network capacity**. When there is a capacity issue impacting the user experience, it may be necessary to increase network capacity by adding radio resources. With an unlicensed network, there is the same need to scale the network and add radio resources as the number of endpoints increases, which is a cost to the owner in order to maintain the network. In doing this, the capacity limitations for the various technologies must be considered. For example, a typical Wi-Fi **access point (AP)** supports around 200 simultaneous users while a Bluetooth gateway typically supports fewer than 50 endpoints. In both licensed and unlicensed networks, the network capacity is dependent on both the number of users and the data throughput for each user, so this ongoing network capacity monitoring is a critical aspect of network management and must be considered in your enterprise IoT solution planning.

Range

The ranges of the various wireless technologies shown in *Figure 3.4* are primarily dependent on the power, frequency, and data rate of the standardized technologies. Since there are regulatory rules for the power and frequency of technologies in both licensed and unlicensed spectrum globally, changes to these parameters must comply with these rules. In the US, this is regulated and managed by the FCC.

The ranges are also affected by interference and obstacles such as walls, so they should be considered general guidelines for the network planning of your enterprise IoT solution. Although there are some exceptions for outdoor deployments and special long-range, low data-rate modes as in Bluetooth 4 or Wi-Fi HaLow, the range of the Bluetooth, Wi-Fi, and LR-WPAN technologies is generally up to 100 meters, while LoRa and cellular LPWA technologies are up to 10,000 meters. These range guidelines, along with required network capacity, are important when deciding the coverage area of your enterprise IoT solution.

Power consumption

Power consumption is a critical factor with enterprise IoT solutions that use battery-powered devices such as long-term asset tracking and remote sensor applications. The required device reporting interval, battery size, wireless technology, and the coexistence of multiple technologies (such as having GPS or Wi-Fi with a cellular device to resolve locations) all impact the IoT device battery life. Although the power consumption of the IoT WLAN/LPWAN technology when fully active and transmitting is important, the power consumption when the device is inactive and in its *sleep* mode is more critical, as the device generally spends over 95% of its time sleeping. To conserve power in some IoT applications, the device is simply powered *off* when not transmitting and awakened when needed. This methodology has the drawback of not being able to reach the device when it is unpowered and having to wait for it to wake up. The more common approach used in most enterprise IoT solutions is to put the device in a deep *sleep* mode with the ability to remotely wake it when needed. As will be discussed in *Chapter 4, Leveraging Cellular IoT Technologies*, cellular LPWA technologies such as LTE-M have specific features to allow the device to *sleep* while still being registered on the network.

Data throughput

The data throughput requirement of your enterprise IoT solution is another critical factor in selecting the best wireless technology. If the IoT data being sent is just sensor or location data, any of the technologies discussed in this chapter will suffice, and the low data-rate technologies will make more sense from a cost and power consumption perspective. If this data from hundreds of endpoints is being aggregated on a gateway and sent wirelessly to an enterprise cloud application, this wireless backhaul technology should be cellular or a higher-throughput Wi-Fi protocol. As well, if there is a need to perform device **firmware-over-the-air** (**FOTA**) updates, which can be several megabytes, either cellular or Wi-Fi is a better choice than LoRa or other WLAN technologies. Finally, if the enterprise IoT solution requires high-definition video and/or **Voice over IP** (**VOIP**), cellular or Wi-Fi is typically the best technology choice.

Module cost

The radio chipset or module cost associated with each of the IoT wireless technologies discussed in this chapter is only important if you are planning to design and build your own IoT device. A module is a **printed circuit board** (**PCB**) with all the integrated radio chipset components, including baseband, application processor, and RF. A device design using a discrete chipset as opposed to a module has a lower **bill of material** (**BOM**) cost but is more complex to design and test. As a general guideline, Bluetooth modules in volumes of around 50,000 are less than $5 while the Wi-Fi, LoRa, LR-WPAN, and cellular LPWA modules are between $6 and $15 depending on features.

In summary, it is important to understand the trade-offs between unlicensed and licensed technologies and scope the best technology for your enterprise IoT solution based on these features—especially coverage, power consumption, and data throughput—to ensure long-term success. In the next section, we will explore the common IoT use cases associated with the technologies reviewed here.

Wireless technology use cases

To provide some context for how the IoT wireless technologies discussed previously can be applied to your enterprise IoT solution, we will discuss the common IoT use cases associated with each technology. Although there is overlap where multiple technologies can be applied to the same IoT use cases, our goal is to present how the technologies are typically used in the IoT market today.

Bluetooth/BLE

As discussed earlier, Bluetooth is embedded in almost all consumer electronic devices such as smartphones, wearables, speakers, headphones, and televisions. With over 4.7 billion Bluetooth devices shipped in 2021 (*Statista 2022* `https://www.statista.com/statistics/1220933/` `global-bluetooth-device-shipment-forecast/`), Bluetooth is one of the most widely used wireless technologies deployed globally. Everyone is probably most familiar with Bluetooth speakers and headphones, which are paired with smartphones and televisions, but there is a myriad of use cases for Bluetooth including healthcare, wearables, and home automation, especially with BLE introduced in Bluetooth 4.0 and Bluetooth Mesh introduced in Bluetooth 5.0.

While classic Bluetooth is a **point-to-point (P2P)** wireless pairing technology, BLE specifies a point-to-multipoint/broadcast model that enables data to be broadcast to many receivers simultaneously. This broadcast model with its very low power enables several interesting IoT solutions. For example, BLE navigation beacons are used to broadcast location parameter data and enable tracking of almost anything from assets/inventory in a warehouse to people. BLE beacons can be used to enable indoor location with an accuracy of less than 5 meters.

In addition to location data, coin-cell battery-powered BLE beacons with sensors are used to broadcast common sensor data such as temperature, humidity, gas detection, leak detection, and vibration in IIoT solutions. With its low power, BLE sensor devices could even be powered exclusively by energy harvesting from solar or thermoelectric sources, saving the need for battery replacement. In healthcare **remote patient monitoring (RPM)** IoT applications discussed in *Chapter 1, Transforming to an IoT Business*, BLE is increasingly being used with blood pressure monitors, glucose meters, weight scales, and heart monitors, where the BLE sensor connects to health wearables, smartphones or other dedicated gateways with a WWAN backhaul connection.

The Bluetooth mesh profile introduced in Bluetooth 5.0 was designed specifically for IoT smart lighting applications. In practice, the building or smart city lights can each be a connected Bluetooth mesh node that can be securely controlled as a group or individually. Lighting-control IoT solutions can not only improve energy efficiency but also customer value, and this is a strong market. With the total revenue of the global smart lighting control system market at $56.2 billion in 2021, the total global smart lighting control systems market is estimated to reach $108.5 billion by 2028 (*Vantage Market Research, 2022*). With their prevalence in the market, low cost, and low power, Bluetooth and BLE technologies will continue to be critical in many enterprise IoT solutions, and the ecosystem of Bluetooth beacons, tracking tags, and sensing tags will continue to increase with the growth of IoT.

Wi-Fi

With over 3 billion Wi-Fi-enabled devices shipped globally each year (*Global Wi-Fi Enabled Devices Shipment Forecast, 2020-2024,* `https://www.researchandmarkets.com/reports/5135535/` `global-wi-fi-enabled-devices-shipment-forecast,` *Research and Markets. 1 July 2020. Retrieved 23 November 2020*), Wi-Fi is one of the most widely used wireless technologies in the world, with in-building coverage almost ubiquitous. It is commonly used in conjunction with both wired Ethernet and cellular WAN technologies in several enterprise IoT solutions, including WWAN backup/failover service, vehicle services, location services, and high-bandwidth applications such as connected security cameras. Given the existing infrastructure of Wi-Fi in homes and buildings, home and building automation applications are obvious IoT applications. One drawback to using Wi-Fi in IoT solutions is the higher power consumption compared to the other wireless technologies, which means the IoT devices typically need external power. This is the impetus for the ratification of HaLow discussed earlier, which addresses range and power concerns for IoT applications.

Wi-Fi is the perfect complementary low-latency, high-bandwidth technology with cellular LTE and emerging 5G technologies, allowing seamless WLAN/WWAN coverage in both consumer and enterprise markets. In an increasing number of enterprise IoT solutions, both Wi-Fi and cellular are deployed in tandem to support a variety of applications where IoT data could be sent to the enterprise cloud application using Wi-Fi without the data costs of cellular LTE or 5G.

As discussed in *Chapter 1, Transforming to an IoT Business*, Wi-Fi is also commonly used in enterprise IoT solutions requiring indoor location services where GPS is not available. In this case, the IoT device reports the Wi-Fi APs that it "sees" and a cloud-based location service such as Google or Azure provides a crowd-sourced "lookup" of the AP location to the enterprise IoT cloud application.

Given the ubiquity of existing Wi-Fi infrastructure and the recent ratification of HaLow, Wi-Fi will continue to be a critical wireless technology in enterprise IoT solutions.

LoRa

LoRa and LoRaWAN LPWA wireless technologies were designed specifically for long-range and low-power IoT applications in both rural and indoor environments, covering nearly all IoT markets including smart cities, IIoT, smart homes/buildings, and smart utilities. Although LoRa lacks the strong ecosystem and ubiquity of Bluetooth and Wi-Fi, it has the advantage of generally lower device cost and much better range/coverage. Other limitations of LoRa technology are the low data throughput of around 50 Kbps and the requirement for a WAN gateway device to send the LoRa device data to the enterprise cloud applications. Even with these limitations, the LoRa and LoRaWAN device market size is projected to reach $6.2 billion by 2026 (*Industry ARC 2022*, `https://www.industryarc.com/Report/19424/lora-and-lorawan-devices-market.html`), driven primarily by enterprise IoT solutions in smart factories, smart cities, smart agriculture, and utilities.

A single LoRa gateway can support thousands of sensor endpoints, which makes LoRa ideal for enterprise IoT solutions with a high number of low data-rate sensor nodes, as in smart agriculture crop monitoring, water/electric metering, and smart factory machine monitoring. There are many low data-rate IoT sensor applications in all IoT markets where LoRa could be used with a wired or wireless WAN technology providing the backhaul to the enterprise cloud application. As such, LoRa and LoRaWAN will have an important role in enabling the billions of connected IoT devices over the next 5 years.

LR-WPAN

As discussed earlier, the LR-WPAN IEEE *802.15* protocol is the foundation for the ZigBee, WirelessHART, and Thread mesh networking standards, which are primarily used in low data-rate home and building

IoT networks for connected lighting, appliances, thermostats, and access control. As with Bluetooth, LR-WPAN devices are low-power and typically battery-powered with long battery life. LR-WPAN networks use a mesh topology and are self-healing, which means all devices on a network can communicate with each other. As with other unlicensed technologies, all LR-WPAN technologies require a wireless or wired WAN gateway to backhaul the data to the enterprise cloud application.

Matter is a new connectivity standard being developed by the **Connectivity Standards Alliance (CSA)** and backed by well-known brands such as Amazon, Apple, and Google primarily for smart home connectivity under a single application framework. Matter works over both Wi-Fi and Thread network standards and, with its strong industry backing, will be the critical smart home IoT connectivity standard enabling a stronger interoperable ecosystem of smart home devices and applications.

Overall, the LR-WPAN technologies will be critical IoT technologies in the smart home, and with strong industry support, Matter should enable the seamless connectivity of home appliances that have traditionally been isolated from each other.

With our review of unlicensed and unlicensed spectrum and WLAN/WWAN technologies, along with our discussion of common use cases with these technologies, let us now review some best practices in deploying IoT wireless technologies to enable more robust enterprise IoT solutions.

Best practices in deploying IoT wireless technologies

When selecting the IoT wireless technology for your enterprise IoT solution, whether it is one of the unlicensed technologies discussed earlier or a licensed cellular LPWA technology, there are five best practices to follow.

1. Consider the network coverage required

Coverage and range are critical considerations when selecting a wireless technology. If the solution is indoors only, any of the short-range WLAN technologies could work well with a WAN gateway for data backhaul; however, if the solution requires both indoor and outdoor coverage, an LPWA technology such as LoRa or cellular may be the best choice. The required coverage area will also determine how many WLAN/WPAN access points and WAN gateways are required to support the solution.

2. Consider the long-term network maintenance/support required

Any wireless network, regardless of whether it is licensed or unlicensed, requires long-term maintenance and support to minimize downtime and maintain quality of service. The wireless network includes not only the hardware or physical IT assets but also the software and cloud access, and all these components need to be managed in terms of network performance, cybersecurity, regular updates, and scalability, which brings us to our third best practice.

3. Leverage existing network infrastructure

Given the capital and operational cost of building and maintaining a private wireless network, it is best to leverage existing public wireless network infrastructure when possible. This is primarily the public cellular LTE and 5G networks, but it could also be public unlicensed LoRa or Wi-Fi networks. Some businesses, especially in the utility and manufacturing markets, prefer to own their enterprise IoT network and frequency spectrum to deliver connectivity to areas not served well by public networks. As a result, there are companies offering unlicensed, private wireless networks (such as Anterix), and many of the public cellular network carriers (such as AT&T and Verizon) also offer private networks as well as hybrid public/private networks to their customers. We will cover private networks in *Chapter 10, Looking at the Road Ahead*.

4. Plan for device management

One area that is overlooked in selecting a wireless technology and developing an enterprise IoT solution is device management. Device management includes everything from device onboarding, configuration, and provisioning to device health and firmware updates in the field. This topic will be covered in more detail in *Chapter 9, Managing the Cellular IoT Solution Life Cycle*. The wireless network needs to be able to support device management functions in terms of adequate data throughput for firmware/software **over-the-air** (**OTA**) updates and a mechanism to *wake up* a sleeping device for health updates and diagnostics. Many enterprise IoT solutions utilize a dedicated device management cloud application specifically for IoT device support, so the wireless network needs to be able to support access to this device management server.

5. Plan on network scaling as the solution grows

Lastly, it is important to plan for network growth as more devices are added, and there is a need to handle the increased network traffic/congestion while still maintaining good data throughput and latency in the enterprise IoT solution. This is implicitly done in public cellular networks but needs to be planned with other public/private unlicensed networks. This will likely require adding network radio resources such as APs and higher bandwidth WAN gateways.

Summary

The goal of this chapter was to provide a technical overview of unlicensed versus licensed technologies and the most prevalent IoT wireless technologies, as well as the associated use cases as a foundation for your enterprise IoT solution. You should have a good context for the trade-offs between licensed and unlicensed wireless technologies and the best practices in selecting and deploying IoT wireless technologies as part of your enterprise IoT solution. In *Chapter 4, Leveraging Cellular IoT Technologies*, we will cover the primary cellular WWAN IoT technologies, which are **Long Term Evolution** (**LTE**), **Category M** (**LTE Cat M**) and **Narrow Band IoT** (**NB-IoT**). The next chapter on cellular LPWA will provide the context on how cellular LPWA is unique to IoT in specific features and use cases and how it is different from standard LTE and 5G technologies. In the rest of this book, we will focus on cellular LTE and 5G technologies and architectures in the context of enterprise IoT solutions, including security, life cycle management, IoT solution case studies, and emerging IoT trends.

Part 2: Deep Dive into Cellular IoT Solutions

With a good understanding of the IoT markets and underlying technologies, we will now review the cellular IoT technologies, including LTE and 5G, as well as cellular IoT devices and network architectures. We will conclude with a couple of real-world cellular IoT solution implementation case studies to bring it all together.

This part contains the following chapters:

- *Chapter 4, Leveraging Cellular IoT Technologies*
- *Chapter 5, Validating 5G with IoT*
- *Chapter 6, Reviewing Cellular IoT Devices with Use Cases*
- *Chapter 7, Securing the Internet of Things*
- *Chapter 8, Implementing an IoT Solution with Case Studies*

4

Leveraging Cellular IoT Technologies

In this chapter, we will provide an overview of **Long-Term Evolution** (**LTE**) cellular IoT technologies with 5G, covered in *Chapter 5, Validating 5G with IoT*. There are specific LPWA LTE technologies called **LTE Category M** (which we will refer to as **LTE-M**) and **Narrow Band IoT** (**NB-IoT**), which are primarily used in low-power enterprise cellular IoT solutions. Both of these LTE LPWA technologies will be covered in this chapter. As we will describe, there are unique features of LTE-M and NB-IoT that have helped drive the global cellular IoT device market to reach 2 billion in 1H21 (IoT Analytics, 2021) with the number of cellular IoT connections expected to reach 3.5 billion in 2023 (Ericsson 2022). While in 2020, unlicensed LPWA technologies (such as LoRa) led licensed LTE LPWA connections with a 53% market share, one year later, LTE LPWA connections led with a 54% share (IoT Analytics, 2021).

In this chapter, we will provide a technical overview of the primary LTE cellular technologies with a focus on the LPWA LTE-M and NB-IoT technologies. Although 2G and 3G networks are still deployed in many countries, these technologies are progressively being sunsetted, with LTE and 5G technologies being the network technologies going forward for the next 20+ years. Since this chapter is focused on LTE and LTE LPWA technologies, we will cover the following topics in this chapter:

- The evolution of cellular technology
- LTE technologies
- LTE LPWA design and best practices
- LTE LPWA use cases

In this chapter, you will learn about the technical details and features of LTE cellular technology in general and, more specifically, LTE LPWA. You will also learn about the LTE LPWA IoT use cases as well as specific design and deployment considerations in the context of an enterprise IoT solution.

Before we start our discussion of LTE technologies, let's go through a brief history of cellular technology.

The evolution of cellular technology

In this section, let's review a brief history of cellular technology, beginning with the introduction of analog (1G) cellular phones in 1979 and continuing up to modern smartphones in the 2020s, as shown in the following chart.

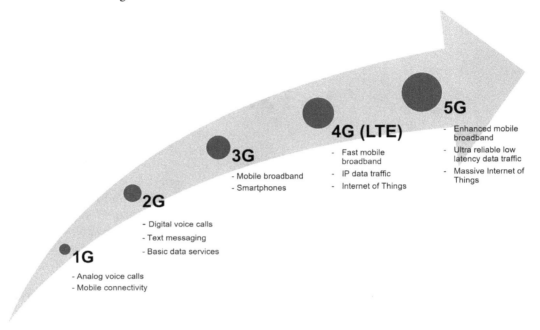

Figure 4.1 – The evolution of cellular technology to 5G

Cellular phones have driven the evolution of cellular network technology globally, starting with analog voice calls in the 1980s, transitioning to digital voice calls and basic data services with 2G and 3G technologies over the following 20 years, and eventually true smartphones with high bandwidth **Internet Protocol (IP)** mobile broadband applications, since roughly 2010. In this network evolution, there were early 2G and 3G devices that were used in **Machine to Machine (M2M)** solutions before it was known as IoT. For example, there were 2G and 3G asset tracking and remote monitoring IoT solutions using integrated cellular modules/modems for connectivity. One of the main disadvantages of using 2G/3G technologies for IoT solutions is the high power consumption of 2G/3G devices compared to other **Wireless Local Area Network (WLAN)** and modern LTE devices. Since the 2G/3G networks were not specifically designed for LPWA IoT solutions, these solutions tended to be inefficient with regard to power, so the device would typically turn *off* the cellular modem when it was not needed to save battery power. The cumulative power consumption required for the device to continually reattach to the network when the device is turned back *on* can be significant over the life of the device. Moreover, by not staying connected, two-way, real-time communication with devices is difficult. Finally, the cost of the higher bandwidth 3G modules and devices that supported data throughputs of up to 14 Mbps was high compared to other WLAN technologies, which created a high barrier of entry for businesses developing an enterprise IoT solution. The advent of 4G LTE in 2010 and, specifically, LTE LPWA in 2016 has filled the gap that was being served by unlicensed LPWA technologies (such as LoRa) and enabled lower-power and cheaper cellular devices, which have been the impetus for the growth of the cellular IoT market, enabling many enterprise IoT solutions.

The standards-based network evolution of cellular technology from 3G to 5G has been managed by the **3rd Generation Partnership Project (3GPP)**, which is a communications industry collaboration of over 400 members that ensures an ecosystem of interoperable cellular devices across suppliers and cellular carriers. As shown in the timeline in *Figure 4.2*, the 3GPP released LTE Cat-1 in Release 8 in 2009, followed by the LTE **Machine Type Communication (MTC)** standard in Release 13 in 2016, which includes the LTE-M and NB-IoT LPWA cellular technologies. Both LTE Cat-1 and MTC (LTE-M and NB-IoT) are the main cellular technologies used in enterprise IoT solutions today. Although the LTE 3GPP standardization releases will end in the next 5 years, all LTE and 5G networks are based on 3GPP standards, and most LTE cellular carriers plan to support LTE network technologies through to at least 2030, with 5G technology being deployed in parallel. The 3GPP standards body has created economies of scale for cellular IoT solutions, due to a large number of companies in the LTE ecosystem (as discussed later in this chapter) that have continued to implement the 3GPP cellular standards over the past 20+ years.

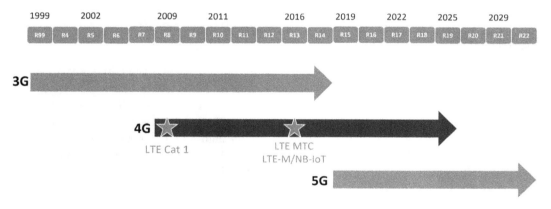

Figure 4.2 – A 3GPP release timeline

Now that we have seen how cellular technology has evolved over time, let's begin our discussion on the LTE technologies that are the basis for most enterprise IoT cellular solutions.

LTE technologies

As of 2022, LTE based on at least the 3GPP Release 8 standard has been deployed in licensed spectrum in over 170 countries, with over 530 cellular carriers; however, the LTE MTC (including LTE-M and NB-IoT) part of the 3GPP Release 13 standard has only been deployed in 74 countries, with 170 cellular carriers. Unlike standard LTE (LTE Cat 1), the cellular LPWA LTE-M and NB-IoT technologies are not as ubiquitous globally, with some countries deploying only LTE-M, only NB-IoT, or both. This is a key consideration for global enterprise IoT solutions, which we will discuss later in this chapter when we review LTE LPWA best practices and network coverage.

We covered the high-level LTE network architecture in *Chapter 2, Understanding IoT Devices and Architectures*, including how **user equipment** (**UE**) or IoT devices access the LTE network via the LTE eNodeB **Radio Access Network** (**RAN**). We also reviewed the elements of the LTE **Evolved Packet Core** (**EPC**), which validates the UE and provides access to the **Packet Data Network** (**PDN**) for the IP to connect to the IoT application server. In this section, we will go into more detail about the call flow to establish a data connection between the UE and IoT application server and discuss in more detail the critical **Subscriber Identity Module** (**SIM**) and **Access Point Names** (**APNs**), used by all cellular IoT devices to gain network access. We will also review the LTE radio technology and licensed LTE frequency bands used globally. We will conclude with an overview of the LTE technology ecosystem players supporting cellular IoT solutions and some general LTE IoT use cases. Let's start with a review of how cellular IoT devices access an LTE network.

LTE network

LTE is the successor technology of earlier 3GPP standards known as the **Universal Mobile Telecommunication System** (**UMTS**) and better known as *3G*. LTE is also known as **Evolved UMTS Terrestrial Radio Access** (**E-UTRA**) and **Evolved UMTS Terrestrial Radio Access Network** (**E-UTRAN**). Referring back to the LTE network diagram in *Figure 2.5* of *Chapter 2, Understanding IoT Devices and Architectures*, shown again in *Figure 4.3*, the high-level LTE network architecture is comprised of the UE, the E-UTRAN **Radio Access Network** (**RAN**) with cellular base stations called eNodeBs or eNBs, and the **Evolved Packet Core** (**EPC**), which manages the UE on the network.

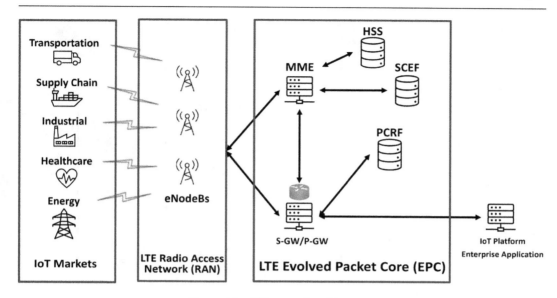

Figure 4.3 – LTE network architecture

Each UE on the LTE network typically communicates with the nearest eNB cellular radio tower (base station), which is usually less than 10 KM from the UE, and the eNBs control the mobile or stationary UE devices in their cell. This communication includes both the data payload to be sent across the network, as well as the 3GPP-compliant LTE signaling protocols. Per the 3GPP standard for LTE, there are several signaling handshakes between the UE, eNB, and the elements of the EPC in order for the UE to set up a data session with the enterprise IoT application. We will now review this call flow.

LTE network call flow

As shown in *Figure 4.4*, there are a number of signaling and authentication exchanges between the UE, eNB, and EPC elements, which includes the **Mobility Management Entity (MME)**, **Home Subscriber Server (HSS)**, **Serving Gateway (SGW)**, and **Packet Data Network Gateway (PGW)** to establish an IP data session with the enterprise IoT application server.

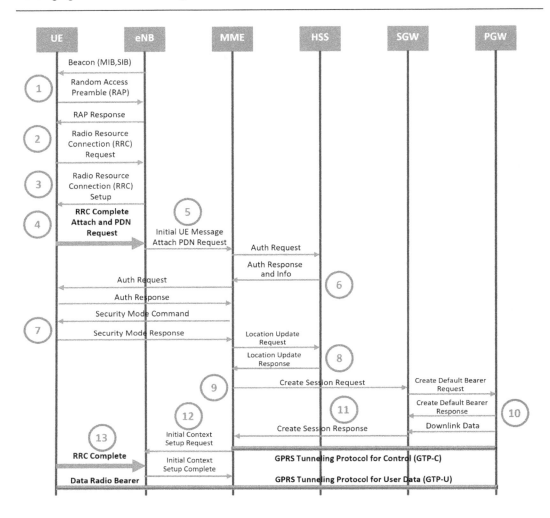

Figure 4.4 – LTE network call flow

To begin the call flow, the UE looks for the **Master Information Block (MIB)** and **System Information Block (SIB)** beacons from the eNB, which allows the UE to find and synchronize with the LTE network. After this is completed, the call flow shown in *Figure 4.4* to establish an end-to-end data session is as follows:

1. The UE sends a **Random Access Preamble (RAP)** message to the eNB to achieve UL synch with the network.

2. The RAP response from the eNB allows the UE to send a **Radio Resource Connection (RRC)** request to the eNB, which contains the UE identity information and establishment cause for the RRC.

3. The eNB then sends a DL message to the UE in order to create what is called a **Signaling Radio Bearer (SRB)**, used for future messages.

4. The UE sends a message to the eNB, indicating that the RRC has been completed and initiates the **Packet Data Network (PDN)** attach request that includes the **Access Point Name (APN)**, which we will cover in the next section.

5. The eNB then sends its first message to the MME that contains this initial UE message, which includes the request for PDN connectivity with the APN.

6. In turn, the MME will perform a security authentication with the HSS that is sent back to the UE for an authentication response.

7. The MME then sends a security mode command to the UE with encryption algorithms, and the UE responds with its ciphering protection.

8. The MME sends a location update request to the HSS that contains information on the PDN subscription details.

9. Upon receipt of the location update response from the HSS, the MME initiates a session request message to the SGW to create a **GPRS Tunneling Protocol (GTP)** tunnel.

10. In turn, the SGW will send this request to the PGW to add the UE to the **Evolved Packet System (EPS)** bearer table, and the PGW response to the SGW will contain the PDN information and EPS bearer identity with downlink data.

11. The SGW then provides a create session response to the MME, indicating that the GTP tunnel is established.

12. The MME then sends the eNB the initial context setup message for the RRC, and the eNB will authenticate the RRC security mode of the UE with the new RRC security keys.

13. Finally, the UE acknowledges the RRC attachment is complete on the new default radio bearer, and the UE can exchange data with the enterprise IoT application.

This LTE call flow is intrinsic to all LTE-compliant chipsets and modules, so the integrated LTE IoT device with these chipsets/modules inherits this functionality. It is important to understand this LTE call flow to troubleshoot potential issues with network connectivity between the cellular IoT device and enterprise application. The most common issue with IoT devices not being able to connect is a mismatch between the network-provisioned APN and the configured APN on the IoT device. Let us now review the LTE APN in more detail.

LTE APN

As part of the call flow when accessing the LTE network, the UE will normally use a carrier-specific APN configured on a device that is used by the EPC (LTE core network) to identify the PDN IP network that the UE wants to communicate with. An APN is essentially an identifier between the cellular network and the internet. As reviewed in *Chapter 2, Understanding IoT Devices and Architectures*, the APN and associated PDN could be public or private. A public APN is shared by many subscribing

IoT devices (UEs) and is more exposed to open internet security vulnerabilities, such as malware and botnet attacks, which will be discussed in *Chapter 7, Securing the Internet of Things*. While most APNs are used publicly, a private APN enables a higher level of privacy and security compared to public APNs by allowing for a secure **Virtual Private Network** (**VPN**) connection to a specific enterprise network. Moreover, a private APN also allows for static or dynamic IP addressing, traffic management rules, and higher network service levels. Using a private APN is recommended for all enterprise IoT solutions. A critical component for a device to gain access to any LTE network is the **SIM** (short for **Subscriber Identity Module**), which we will now cover.

LTE SIM

Each UE or device on an LTE network has an integrated **Universal Integrated Circuit Card** (**UICC**), also known as a SIM card, that identifies the UE on the network and ensures the privacy and security of the data sent on the network. The SIM has the UE carrier account profile, and the SIM provisioning in the core HSS includes the carrier-specific APN, as discussed in the previous section on the LTE network. SIM provisioning allows specific network features such as voice calls, text messaging, data usage, subscription billing, and network roaming with other cellular carriers. Each SIM is internationally identified by its unique **Integrated Circuit Card Identifier** (**ICCID**) number, which includes the unique **International Mobile Subscriber Identity** (**IMSI**) of the cellular carrier (such as AT&T, Vodafone, and Verizon). In the context of an enterprise IoT solution, it is important to ensure the SIM-provisioned features, especially with regard to international roaming, match the IoT deployment scenarios, as SIM provisioning defines where a device can roam geographically.

As cellular technology has evolved over the past 30+ years, so have SIMs, which are standardized by the **Global System for Mobile Communications** (**GSMA**) association. As shown in *Figure 4.5*, SIMs have four primary sizes or form factors, with 2FF (mini SIM), 3FF (micro SIM), and 4FF (nano SIM) being the standard plastic exchangeable SIMs that are inserted in the SIM holder of a UE device. The MFF2 SIM introduced in 2016 is an embedded SIM that is soldered to the **Printed Circuit Board** (**PCB**) in the UE, and it has the advantages of a higher operating temperature and being much smaller and much more resilient to device vibrations, which can cause issues with the SIM's electrical connections.

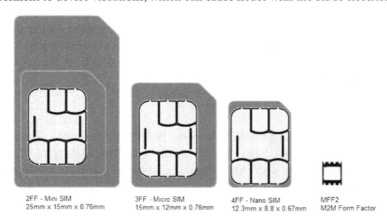

2FF - Mini SIM
25mm x 15mm x 0.76mm

3FF - Micro SIM
15mm x 12mm x 0.76mm

4FF - Nano SIM
12.3mm x 8.8 x 0.67mm

MFF2
M2M Form Factor

Figure 4.5 – SIM sizes

In addition to these standard SIMs, the **Integrated SIM (iSIM)** or **Integrated Universal Integrated Circuit Card (iUICC)**, introduced in 2020, is the most advanced version of a SIM. The iSIM is embedded within a **Tamper Resistant Element (TRE)** on a device's cellular **System on Chip (SoC)**, eliminating the need for any discrete SIM hardware on the device. In other words, the iSIM is embedded in the LTE chipset on the device. Like standard SIMs, the iSIM securely stores the cellular carrier profile and can be remotely provisioned without physically replacing the SIM. Using an iSIM is a strong trend with cellular IoT solutions, as it reduces the overall size and complexity of devices. Moreover, an iSIM saves device power by using the processor of the device SoC without the need for a separate processor component, such as a **Microcontroller Unit (MCU)**. All recent versions of standard SIMs and iSIMs can also be **Embedded Universal Integrated Circuit Card (eUICC)**-compliant, which allows remote carrier provisioning of the SIM profile. This means the **Mobile Network Operator (MNO)** or **Mobile Virtual Network Operator (MVNO)** carrier can be switched without physically replacing the SIM in the UE. This ability to switch carriers requires the MNO or MVNO to have a **Subscription Management (SM)** platform that includes **Subscription Management Data Preparation (SM-DP)** and **Subscription Management Secure Routing (SM-SR)** components. Both the SM-DP and SM-SR components are required to remotely provision an eUICC SIM. The obvious advantage of eUICC is that an enterprise cellular IoT solution need not be locked into a specific MNO or MVNO cellular carrier. This allows a user to dynamically select cellular carriers even after devices are deployed in the field, provided the carrier has the infrastructure to support eUICC remote provisioning. We will cover the important growing trends of iSIM/iUICC and eUICC in more detail in *Chapter 10, Looking at the Road Ahead*.

Now that we have introduced the high-level LTE network and the APNs/SIMs enabling network connectivity, we will now review the LTE radio technology. This will be the basis for our discussion of specific LTE LPWA technologies later in this chapter.

LTE radio technology

As discussed in the introduction on the evolution of cellular technology, the impetus for LTE standardization was the emergence of new smartphones and high-bandwidth streaming multimedia applications. As such, LTE was designed to be a more efficient, high-data-rate, and low-latency IP-based system, with flexible bandwidth deployments from 1.4 MHz to 20 MHz (the higher the frequency bandwidth, the higher the supported data rate). To achieve higher data rates and spectral efficiency, LTE was designed to support **Multiple Input Multiple Output (MIMO)**, which uses multiple transmitters and receivers to transfer more data, especially in the **Downlink (DL)** from the cell tower to the device, also called the UE. In *Table 4.1*, the maximum DL and UL data rates with the associated 3GPP release for the various LTE UE categories are shown. The UE categories, **NB1** and **M1**, refer to the LTE LPWA technologies NB-IoT and LTE-M respectively, which we will discuss in more detail later in this chapter. As shown in the table, LTE can support theoretical DL data rates from 0.68 Mbps with NB-IoT up to 25 Gbps with LTE Category 17. This wide range of potential UE categories allows for a broad range of UE devices in the LTE ecosystem, from high-speed routers and smartphones/tablets to low-cost, low-power sensors and IoT devices, using LTE-M and NB-IoT, which we will cover in more detail later in this chapter.

UE category	Maximum theoretical DL data rate (Mbps)	Maximum number of DL MIMO layers	Maximum theoretical UL data rate (Mbps)	3GPP release
NB1	0.68	1	1	13
M1	1	1	1	13
0	1	1	1	12
1	10.3	1	5.2	8
2	51	2	25.5	8
3	102	2	51	8
4	150.8	2	51	8
5	299.6	4	75.4	8
6	301.5	2 or 4	51	10
7	301.5	2 or 4	102	10
8	2,998.6	8	1,497.8	10
9	452.2	2 or 4	51	11
10	452.2	2 or 4	102	11
11	603	2 or 4	51	11
12	603	2 or 4	102	11
13	391.7	2 or 4	150.8	12
14	3917	8	9,585	12
15	750	2 or 4	226	12
16	979	2 or 4	N/A	12
17	25,065	8	N/A	13
18	1,174	2, 4, or 8	N/A	13
19	1,566	2, 4, or 8	N/A	13

Table 4.1 – LTE UE category definitions

The data in *Table 4.1* is taken from `https://www.everythingrf.com/community/lte-frequency-bands`.

LTE can be deployed on various operating frequency bands, as shown in *Table 4.2*, depending on the country and cellular carrier, so when deploying an IoT solution globally, it is important to ensure the IoT device and cellular carrier supports not only the LTE technology category shown in *Table 4.1* but also the LTE frequency bands in each country where they are deployed. Provided the device also supports 2G/3G technologies, LTE supports seamless handover between the legacy 2G and 3G networks which is important for device interoperability. As shown in *Table 4.2*, LTE can be deployed in both **Frequency Division Duplex (FDD)** and **Time Division Duplex (TDD)** modes, depending on the frequency operating band. In TDD mode, the UL and DL use the same frequency bands but transmit and receive data traffic at different times, whereas, in FDD mode, the UL and DL use different frequency bands to transmit and receive data traffic.

LTE FDD Frequency Bands		
LTE Band Number	**Uplink Band (MHz)**	**Downlink Band (MHz)**
LTE Band 1	1920 - 1980	2110 - 2170
LTE Band 2	1850 - 1910	1930 - 1990
LTE Band 3	1710 - 1785	1805 - 1880
LTE Band 4	1710 - 1755	2110 - 2155
LTE Band 5	824 - 849	869 - 894
LTE Band 6	830 - 840	875 - 885
LTE Band 7	2500 - 2570	2620 - 2690
LTE Band 8	880 - 915	925 - 960
LTE Band 9	1749.9 - 1784.9	1844.9 - 1879.9
LTE Band 10	1710 - 1770	2110 - 2170
LTE Band 11	1427.9 - 1452.9	1475.9 - 1500.9
LTE Band 12	698 - 716	728 - 746
LTE Band 13	777 - 787	746 - 756
LTE Band 14	788 - 798	758 - 768
LTE Band 15	1900 - 1920	2600 - 2620
LTE Band 16	2010 - 2025	2585 - 2600
LTE Band 17	704 - 716	734 - 746
LTE Band 18	815 - 830	860 - 875
LTE Band 19	830 - 845	875 - 890
LTE Band 20	832 - 862	791 - 821
LTE Band 21	1447.9 - 1462.9	1495.5 - 1510.9
LTE Band 22	3410 - 3500	3510 - 3600
LTE Band 23	2000 - 2020	2180 - 2200
LTE Band 24	1625.5 - 1660.5	1525 - 1559
LTE Band 25	1850 - 1915	1930 - 1995
LTE Band 26	814 - 849	859 - 894
LTE Band 27	807 - 824	852 - 869
LTE Band 28	703 - 748	758 - 803
LTE Band 29	-	717 -728
LTE Band 30	2305 - 2315	2350 - 2360

LTE Band 31	452.5 - 457.5	462.5 - 467.5
LTE Band 32	-	1452 - 1496
LTE Band 65	1920 - 2010	2110 - 2200
LTE Band 66	1710 - 1780	2110 - 2200
LTE Band 67	-	738 - 758
LTE Band 68	698 - 728	753 - 783
LTE Band 69	-	2570 - 2620
LTE Band 70	1695 - 1710	1995 - 2020
LTE Band 71	663 - 698	617 - 652
LTE Band 72	451 - 456	461 - 466
LTE Band 73	450 - 455	460 - 465
LTE Band 74	1427 - 1470	1475 - 1518
LTE Band 75	-	1432 - 1517
LTE Band 76	-	1427 - 1432
LTE Band 85	698 - 716	728 - 746
LTE Band 87	410 - 415	420 - 425
LTE Band 88	412 - 417	422 - 427

LTE TDD Bands		
LTE Band Number	**Frequency**	**Bandwidth (MHz)**
LTE Band 33	1900 - 1920 MHz	20
LTE Band 34	2010 - 2025 MHz	15
LTE Band 35	1850 - 1910 MHz	60
LTE Band 36	1930 - 1990 MHz	60
LTE Band 37	1910 - 1930 MHz	20
LTE Band 38	2570 - 2620 MHz	50
LTE Band 39	1880 - 1920 MHz	40
LTE Band 40	2300 - 2400 MHz	100
LTE Band 41	2496 - 2690 MHz	194
LTE Band 42	3400 - 3600 MHz	200
LTE Band 43	3600 - 3800 MHz	200
LTE Band 44	703 - 803 MHz	100
LTE Band 45	1447 – 1467 MHz	20
LTE Band 46	5150 – 5925 MHz	775
LTE Band 47	5855 – 5925 MHz	70
LTE Band 48	3550 – 3700 MHz	150

LTE Band 50	1432 – 1517 MHz	85
LTE Band 51	1427 – 1432 MHz	5
LTE Band 52	3300 – 3400 MHz	100
LTE Band 53	2483.5 – 2495 MHz	11.5

Table 4.2 – LTE-operating frequency bands

Every country has a unique combination of 2G, 3G, LTE, and 5G cellular network deployments, both in terms of operating bands and technologies, with many countries still supporting legacy 2G and 3G technologies in their journey to migrate to LTE and 5G technologies over the coming years.

The data in *Table 4.2* is taken from `https://www.everythingrf.com/community/lte-frequency-bands`.

Especially with LTE LPWA, LTE-M, and NB-IoT technologies playing a critical role in the future of enterprise cellular IoT solutions, it is important to ensure countries where an LTE LPWA solution is planned match with the network operating bands and technology deployed in each country. According to the 3GPP specification from Release 13, the list of supported LTE-M bands is as follows:

Bands 1, 2, 3, 4, 5, 7, 8, 11, 12, 13, 18, 19, 20, 26, 27, 28, 31, 39, 41

According to GSMA, these bands need to be covered for global coverage in North America, Latin America, Europe, and Asia: *Bands 1, 2, 3, 4, 5, 12, 13, 20, 25, 26, 28.*

LTE technology ecosystem

As shown in *Figure 4.6*, there are a number of providers in the LTE technology ecosystem, including the providers of the LTE radio chipset, modules based on the chipsets, devices, SIMs, and network equipment to support end-to-end LTE IoT solutions. We will review cellular IoT devices in detail in *Chapter 6, Reviewing Cellular IoT Devices with Use Cases*, including the chipset and module components, as well as the various device types and use cases.

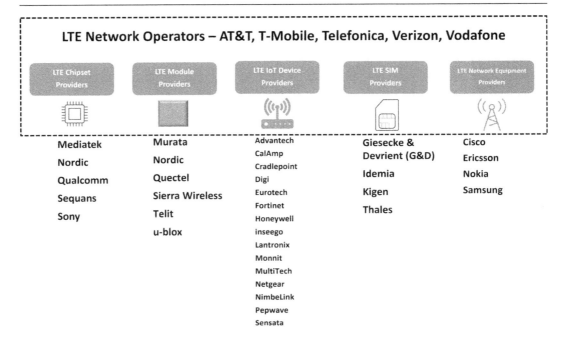

Figure 4.6 – LTE technology ecosystem

As shown in *Figure 4.6*, network operators such as AT&T, T-Mobile, and Verizon bring all of the LTE technology ecosystem together along with additional services, such as IoT solution consulting to enable end-to-end cellular IoT solutions. The LTE network operators certify all of these technology components to ensure the overall performance of their LTE networks. As such, selecting the network operator for your enterprise IoT solution is a critical task early in the design and planning stage. Although we will cover specific LTE LPWA use cases later in this chapter, let us now briefly look at the general LTE IoT use cases.

LTE IoT use cases

As shown in *Figure 4.7*, LTE supports a broad range of IoT use cases, from low-data-rate, higher-latency LPWA use cases such as water metering, environmental monitoring, industrial sensing, and asset tracking, which are mostly battery-powered, to high-data-rate, low-latency applications such as video analytics, connected cars, fleet management, and smart homes.

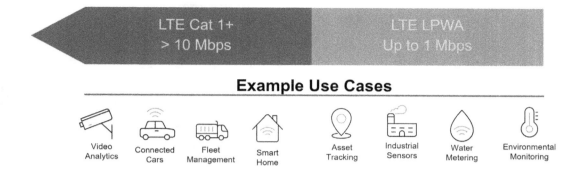

Figure 4.7 – The IoT use cases supported by LTE

While **LTE LPWA** and **Cat 1** are the most common LTE technologies used in cellular IoT solutions due to lower-cost devices and modules, there are many broadband IoT applications that require high data throughputs, such as video surveillance, connected cars, and laptops, which require the higher categories of LTE reviewed earlier. In terms of device types for the higher-category LTE technology solutions, routers and gateways are the most common, as they provide the most flexibility in supporting a wide array of IoT use cases. We will review most of the cellular device types in *Chapter 6, Reviewing Cellular IoT Devices with Use Cases*, with the most common LTE category used with each type.

Now that we have provided an overview of the LTE network, radio technologies, ecosystem, and use cases, we will review the LTE LPWA technologies LTE-M and NB-IoT in more detail, as these technologies have unique features in terms of low power and low cost that enable many enterprise cellular IoT solutions.

LTE LPWA design and best practices

The LTE LPWA technologies LTE-M and NB-IoT specified in 3GPP Release 13 were designed specifically for the IoT markets and typically coexist in a cellular carrier's existing operating frequency bands. The lower cost and power as compared to standard LTE has significantly increased the adoption of LTE LPWA to the point that the number of LTE LPWA connections now tops unlicensed LPWA connections. In this section, we will discuss the unique features of LTE LPWA that enable low-cost and low-power IoT devices. As shown in *Table 4.3*, there is a trade-off primarily in terms of data rates and latency for these features.

Feature	LTE Cat 1	LTE-M	NB-IoT
Bandwidth	20 MHz	1.4 MHz	200 KHz
Receive Antennas	2	1	1
Power Class	23 dBm	20/23 dBm	20/23 dBm
Peak Downlink Rate	10 Mbps	1 Mbps	170 Kbps
Peak Uplink Rate	5 Mbps	1 Mbps	250 Kbps
Voice Support	Yes (VoLTE)	Yes (VoLTE)	No
Mobility Support	Yes	Yes	No
Average Latency	50 to 150 ms	50 to 3000 ms	1600 to 10000 ms

Table 4.3 – A comparison of standard LTE with LTE LPWA

Both LTE-M and NB-IoT require a single receive antenna (versus two for standard LTE), low spectrum bandwidth, and low transmit power. Taken together, these reduced requirements lower a device's cost and complexity. While NB-IoT is exclusively deployed in half-duplex mode, LTE-M can be deployed in either a full-duplex or-half-duplex mode of operation. In half-duplex, the device transmits and receives at different times, which lowers data rates and increases latency, with the benefit of even lower device complexity and cost. For cellular networks that have deployed LTE-M in half-duplex mode, the peak downlink and uplink data rates drop to 375 Kbps from the 1 Mbps shown in *Table 4.3*.

In the context of an enterprise IoT solution, there are some important differences in the capabilities between Cat 1, LTE-M, and NB-IoT. Both LTE Cat 1 and LTE-M can support voice service using **Voice over LTE (VoLTE)** as well as mobility, but NB-IoT supports neither voice nor mobility. In terms of device management in enterprise IoT solutions that will require **Firmware over the Air (FOTA)** device updates, the data rate of LTE-M allows for recommended FOTA file sizes of up to 1 MB, but the low data rates of NB-IoT only allow for recommended FOTA files sizes of less than 250 KB. Although larger FOTA updates are possible with both LTE-M and NB-IoT, there will be a negative impact on the battery life of a device. When using LTE-M/NB-IoT in your enterprise IoT solution, it is important to consider the *data rate, latency, voice, mobility, and FOTA requirements* of the IoT application in the context of these limitations. Overall, LTE LPWA technologies are suitable for most enterprise IoT solutions that are latency-tolerant and don't require high data rates.

There are two primary power-saving features of LTE-M and NB-IoT that bring these LPWA technologies level with the unlicensed LPWA technologies on the market, such as LoRa. These are **Power Saving Mode (PSM)** and **Extended Discontinuous Reception (eDRX)**, as described in the next section.

PSM and eDRX low-power features

Both PSM and eDRX allow the device to stay connected to the network while remaining inactive or in sleep mode for hours up to days, which saves a device's battery power without the need to completely turn off the cellular radio. A good example IoT use case that can benefit from these power-saving features is utility water metering, where the battery-powered IoT water meter device only needs to report to the utility enterprise application monthly; otherwise, it can stay in sleep mode to save battery power. As shown in the power consumption versus time graph in *Figure 4.8*, PSM and eDRX times are requested by the UE and controlled by the T3412 and T3324 network timers respectively. When using these power-saving features, the main consideration is how to handle **Mobile-Terminated (MT)** data, as data can only be received in accordance with the PSM and eDRX network timers.

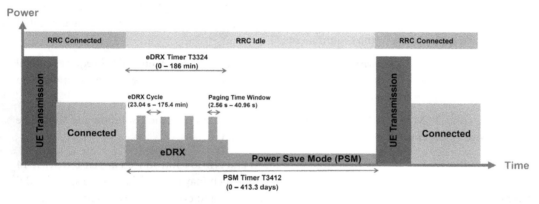

Figure 4.8 – LTE LPWA PSM and eDRX timers

An LTE-M or NB-IoT device can be configured to support PSM, eDRX, or both modes. These features are designed to enable IoT devices to conserve power and potentially achieve a 10-year battery life.

As shown in *Figure 4.8*, there is a **Radio Resource Control (RRC)** idle time, also known as the **Tracking Area Update (TAU)** period, requested by the device and set by the T3412 network timer, which includes the **Power-Saving Mode (PSM)** or hibernation period and the eDRX T3324 timer. The T3412 and T3324 timers are used together to manage the active wake-up time of the LTE-M or NB-IoT IoT device. The maximum time for the T3412 PSM timer is 413 days, and the maximum time for the T3324 eDRX timer is 186 minutes. The T3324 eDRX timer has two sub-timers that set the active eDRX period and eDRX cycle network paging interval when the device is reachable. While the normal LTE paging cycle, also known as the idle mode DRX, is typically either 1.28 seconds or 2.56 seconds, with eDRX, this time can be extended to a maximum of 175 minutes. The actual timer values allowed are dependent on the LTE carrier network configuration. During the PSM period, the monitoring of network paging instances is turned *off* and a device is not reachable for MT data and a **Short Message Service (SMS)**. As such, the device can go into a low-power sleep or hibernation mode until the end of the TAU period, when it is able to send and receive data again. While most LTE networks buffer SMS messages for a few days, SMS delivery to devices using PSM is unreliable and not recommended. With regards to MT data, it is best to implement a *pull* method on the device where the device initiates the transmission/notification to the application server that it is ready to receive MT data. The device can *wake up* at any time during the PSM period to send data, which also resets the T3412 timer. In the case of eDRX, the network normally stores any MT messages and delivers them to the device on the next paging cycle. The size of the network buffer for these MT messages is dependent on the network carrier, so it is important to check with your selected network carrier on the allowed timer values and buffer sizes for both PSM and eDRX. Both PSM and eDRX are critical features of LTE-M and NB-IoT and can enable a 10-year battery life for IoT devices. Both PSM and eDRX can be implemented together to optimize power consumption for an IoT solution. When implementing PSM and/or eDRX in your enterprise IoT solution, here are some best practices to follow:

- For optimum battery life, the recommended ratio of PSM cycle to RRC idle time ((T3412–T3324)/T3412) should be greater than 90%, but this ratio should be selected in the context of your enterprise IoT solution requirements

- Using MT data and SMS with devices using PSM and, to a lesser extent, eDRX requires careful planning for when the device is reachable, so the best approach is to set a device to wake up and notify the application server when it is available to receive messages

- Using either PSM or eDRX increases the latency of the enterprise IoT solution and should be planned in the context of your enterprise IoT solution

There is one additional power-saving feature for LTE-M and NB-IoT introduced in 3GPP Release 14 called **Release Assistance Indicator** (**RAI**). This feature, which may be available on the specific LTE network carrier where the device is deployed, allows a device to indicate that it has no more UL data to send and is not looking for any server response. This allows the device to enter RRC idle early, which saves power in not having to wait for the expiration of the network inactivity timer.

To conclude our discussion of the LTE LPWA LTE-M and NB-IoT technologies, let's explore some general best practices in planning, designing, and deploying LTE LPWA connectivity in your enterprise IoT solution.

LTE LPWA best practices

When selecting licensed LTE LPWA as the best technology for your enterprise IoT solution, there are a few factors to consider in selecting the specific LPWA technology (LTE-M and/or NB-IoT) and network carrier. So, when choosing the best LTE LPWA technology for your enterprise IoT solution, it is important to consider the following technology features with your selected network carrier:

- Network coverage

- Data rate

- Latency

- Mobility

- Voice and SMS

Network coverage

Network carriers normally include LTE-M and/or NB-IoT technologies as part of their standard LTE network deployment, and they have specific LPWA roaming agreements with other global LTE carriers. Most LTE LPWA modules (the modem component integrated into the LPWA device) and integrated IoT devices support both LTE-M and NB-IoT, potentially with 2G as a fallback technology, to enable better global coverage, since neither LTE-M nor NB-IoT is deployed globally.

There is a small cost advantage in selecting single-mode NB-IoT modules/devices, but this is usually outweighed by the need for global roaming. As such, a dual-mode LPWA module/device with 2G fallback provides the best global LPWA coverage; however, standard LTE modules/devices (LTE Cat 1 and higher) will have better overall global coverage. When selecting your LPWA network carrier, it is important to assess whether they have roaming agreements with the carriers in the countries where your IoT solution will be deployed.

Data rate

As discussed earlier in this chapter, LTE-M can support a theoretical peak uplink data rate of 1 Mbps (375 Kbps in half-duplex mode) compared to 250 Kbps for NB-IoT, which is important from the perspective of device power consumption, as the longer a device needs to stay connected to a network, the higher the power consumption. In real-world network deployments, the LTE-M and NB-IoT data rates will be much lower, by over 60%, due to network load and **radio frequency** (**RF**) propagation conditions, so this needs to be factored into the actual data rate requirements of your enterprise IoT solution to support not only device reporting but also FOTA updates for device management. Both LTE-M and NB-IoT also have a unique optional feature called **Coverage Extension** (**CE**) mode, which is an automatic network feature to maintain connectivity in poor coverage areas, such as basements or underground garages. In CE mode, the network automatically repeats messages to the UE up to 32 times when the UE is in poor coverage, to improve UE connectivity and coverage at the cost of much lower data throughput. When a UE is operating in CE mode, the data throughput is significantly lower, and the latency is much higher due to these repetitions. Like the other LPWA features, the implementation and configuration of CE mode are LTE carrier-specific and should be checked with your selected network carrier.

Latency

As stated earlier, the network latency with LTE-M and NB-IoT is much higher than standard LTE (Cat 1+), so if your enterprise IoT solution is sensitive to high latency, LTE LPWA is probably not the best technology choice. For example, in IoT solutions where near real-time device reporting is required, as in critical alarm reporting, it is likely better to select standard LTE, which has a consistent latency of typically less than 150 ms.

Mobility

An important difference between LTE-M and NB-IoT is mobility support. Like standard LTE, LTE-M supports **Connected Mode Mobility** (**CMM**), where devices stay connected while moving across the LTE network, whereas NB-IoT does not. With NB-IoT, a device is forced to re-register with the network when moving to different eNodeB network service areas. For example, LTE-M is commonly used in asset-tracking applications that require the device to roam across multiple LTE eNodeB towers while maintaining connectivity.

Voice and SMS

LTE-M can support SMS and VoLTE, albeit at a lower quality than standard LTE, but NB-IoT does not support voice or SMS, so if voice and/or SMS are needed in your enterprise IoT solution, LTE-M is required. For example, **Mobile Personal Emergency Response System** (**mPERS**) solutions where customers can press a button for assistance require voice, making these low-power solutions a good fit for LTE-M. Not all LTE network carriers support voice or SMS with LTE-M, so it is important to check with your selected network carrier whether voice or SMS is required.

Overall, the specific features of LTE-M and NB-IoT previously reviewed that are required for your enterprise IoT solution should be reviewed with your selected LTE LPWA network carrier. For global IoT solutions, it is also critical to assess the country-specific LTE LPWA technologies, LTE frequency bands, and roaming agreements with your selected network carrier. The major **Mobile Network Operators** (**MNOs**), such as AT&T in the US and Vodafone in Europe, have extensive roaming agreements with global LTE LPWA carriers to enable international roaming for their customers. There are also **Mobile Virtual Network Operators** (**MVNOs**), such as KORE and Aeris, that resell the connectivity services of many global MNOs, providing some flexibility in network carriers for your enterprise IoT solution. The main drawback of using an MVNO is, potentially, customer service and network technical support without a direct relationship with the MNO.

With an understanding of the best practices in deploying an enterprise IoT solution with LTE LPWA, let's now explore some of the IoT LTE LPWA use cases.

LTE LPWA use cases

We reviewed many of the IoT LPWA markets and use cases in *Chapter 1, Transforming to an IoT Business*. So, in this section, we will review both trending and future use cases specific to LTE LPWA, where an IoT sensor device, typically battery-powered, connects directly to an LTE network as opposed to a gateway device with a wired or wireless WAN backhaul. In the context of an enterprise IoT solution, the IoT device normally sends time-series event and sensor data to the enterprise cloud application, which executes an action based on this data (such as predictive maintenance, service calls, meter billing, or remote control of infrastructure) and usually also performs data analytics to identify trends in the data. Some trending LPWA use cases and vertical markets are shown in *Figure 4.9*, with transportation, supply chain logistics, factory monitoring, and utility metering (water, gas, and electric) being the most common.

IoT Market	LPWA Use Cases
Transportation	Fleet management, cargo monitoring, asset tracking, logistics optimization
Supply Chain Logistics	Shipment tracking and monitoring, location of goods in a warehouse
Factory Monitoring	Temperature, pressure, and flow monitoring; predictive maintenance
Utility Smart Metering	Electric, water, and gas metering

Figure 4.9 – Common LPWA use cases

In nearly all LTE LPWA IoT use cases and solutions, a device is battery-powered, and the logistics of battery replacement due to the scale and disparate locations of deployed field devices is not a viable option. As such, proper planning and implementation of the low-power features of LTE LPWA discussed earlier are critical to ensure the enterprise IoT solution has a long life cycle. Moreover, since it is not always clear where devices may be deployed, network coverage is critical to ensure that there are no gaps in device reporting, which is part of the reason licensed LTE LPWA is a better choice over unlicensed LPWA technologies.

In many LTE LPWA IoT use cases with low data rates that don't require mobility or voice, either LTE-M or NB-IoT can be used, as in environmental monitoring and utility metering IoT applications. The small cost difference between single-mode LTE-M and NB-IoT devices/modules combined with the reduced feature set of NB-IoT described earlier makes LTE-M the most popular LPWA technology for nearly all LPWA applications, with NB-IoT typically considered a fallback technology in countries that have not deployed LTE-M. At the time of writing, some of the emerging LTE-M LPWA IoT use cases include the following:

- **Healthcare Remote Patient Monitoring (RPM)**:
  - mPERS
  - Wellness wearables
  - Remote vital sign monitoring
- **Smart cities**:
  - Utility metering
  - Lighting control
  - Energy management

- Climate sustainability (carbon emissions)

- Environmental monitoring (air quality and weather)

- **Smart agriculture**:

  - Irrigation monitoring/control

  - Crop monitoring

  - Livestock monitoring/tracking

All these LPWA IoT applications provide societal benefits as well as new service business models, which make them good candidate applications in an enterprise IoT solution. Given that most LTE network carriers expect to support LTE and LPWA for at least the next 10 years, the network infrastructure and LPWA features should enable many new enterprise IoT solutions based on LPWA in the future.

With that, we have arrived at the end of the chapter. Let's summarize what we covered and how this sets the foundation for the remainder of the book.

Summary

We started our discussion by reviewing the evolution of cellular technology, leading to LTE, which includes the LPWA technologies LTE-M and NB-IoT. Next, we provided an overview of LTE in general to include the network, radio, technology ecosystem, and use cases, which led us to a focused review of LTE LPWA, including the low-power features and best practices in deploying IoT solutions based on LTE LPWA. We concluded with an overview of the LTE LPWA IoT use cases to provide some context for your enterprise IoT solution. Overall, this should give you critical information on how best to design and deploy your IoT solution with LTE technology. In the following six chapters, we will cover the latest 5G technology in the context of IoT and provide more insight into LTE/5G IoT devices, architectures, and security guidelines. We will conclude our review of cellular IoT enterprise solutions with a few real-world case studies, a review of the cellular IoT solution life cycle with some best practices, and look at some emerging IoT technologies, trends, and business models.

Validating 5G with IoT

In this chapter, we will provide an overview of the newest cellular technology, called 5G or 5G **New Radio (NR)**, which has been deployed primarily in the US, China, Japan, and South Korea with 198 commercial 5G networks launched globally as of May 2022 [**Global System for Mobile Communications (GSMA)**, May 2022]. The 5G standard was released in the **3rd Generation Partnership Project (3GPP)** Release 15 in 2018 and, as shown in *Figure 5.1*, was designed to address the three key service areas of mission-critical control with **Ultra Reliable Low Latency Communications (URLLC)**, **enhanced mobile broadband (eMBB)** with high data rates and capacity, and massive IoT or **massive Machine Type Communication (mMTC)** with low complexity and power but with high density:

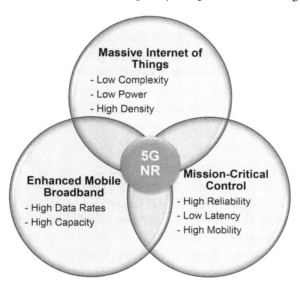

Figure 5.1 – 5G NR services

As shown in *Figure 5.1*, the main tenets of 5G that enable these service areas are low latency of less than 20 ms versus over 100 ms with standard **long-term evolution (LTE)**, high data throughputs of up to 20 Gbps versus less than 2 Gbps with standard LTE, and support of a high density of IoT devices.

Each of the 5G service areas supports different IoT use cases. For example, the mission-critical or URLLC service can support autonomous vehicle and remote surgery applications where low latency and high reliability are critical. The eMBB service can support **augmented reality/virtual reality (AR/VR)** and video analytic applications that require high data rates. Finally, the massive IoT or mMTC service can support a high density of IoT devices on the network, such as water meters or environmental sensors that are low power and latency tolerant. We will review the most common 5G use cases later in this chapter.

In this chapter, we will provide a technical overview of 5G with a focus on the massive IoT aspect. We will also focus on the new innovations of 5G compared to standard LTE, which will enable many new applications. As such, we will cover the following topics in this chapter:

- Overview of 5G

- 5G and IoT

- 5G use cases

- Best practices in 5G solutions

Let's begin with a technical overview of 5G.

Overview of 5G

Let's begin our review of 5G with an overview of the 5G network architecture, including the 5G network deployment options. We will conclude our overview with a review of the 5G radio technology and frequency spectrum as well as **service-based architecture (SBA)** slicing, supporting unique 5G IoT use cases.

5G network architecture

Most of the 5G NR network deployments today use what is called a **Non Standalone (NSA)** architecture where the LTE network **Evolved Packet Core (EPC)** is used for the control plane data, with 5G used for the user plane data. The control plane manages how data packets are forwarded, whereas the data plane actually carries the user traffic. These networks are typically planned to evolve to a **Standalone (SA)** architecture with both user and control plane data on a dedicated 5G core. These network architectures are shown in *Figure 5.2*:

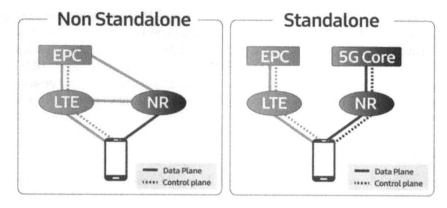

Figure 5.2 – 5G NSA versus SA

Due to existing LTE network infrastructure and cost, operators typically begin their 5G deployments with the NSA architecture, but NSA requires operators to support both LTE and 5G, which is more costly than supporting either one alone. Although both NSA and SA networks use the 5G NR **radio access network** (**RAN**) interface, there are significant differences between the two architectures where SA allows both users and operators to achieve the full advantages of 5G technology. Let us now review the 5G radio technology that enables high throughput and low latency features.

5G radio technology

Compared to standard LTE, 5G uses the same baseband radio technology called **Orthogonal Frequency Division Multiplexing** (**OFDM**), which is also used in the latest Wi-Fi standards. One of the key differences in 5G radio compared to LTE is the requirement for 4 X 4 **MIMO** (short for **Multiple Input Multiple Output**) on **user equipment** (**UE**) devices, which is optional for most LTE devices. A UE device implementing 4 X 4 MIMO would have four physically independent transmit-and-receive **radio frequency** (**RF**) channels with antennas. The 5G base station (called **gNodeB** or **gNB**) can support **massive MIMO** (**mMIMO**) with up to 256 x 256 MIMO. This MIMO system requirement enables a better user experience with higher reliability and higher data throughput. MIMO allows for spatial diversity by sending the same data across different wireless paths or virtual pipelines, which enables the device to resolve the transmission with higher reliability. This helps ensure a more consistent user experience even at cell edges compared to standard LTE. MIMO also allows for RF beamforming where the wireless transmission path can be directed toward specific users, increasing data throughput and reducing interference.

To increase the theoretical data throughput to achieve greater than 20 Gbps, 5G allows for increased channel bandwidths of up to 800 MHz compared to 640 MHz for the most advanced LTE standard called LTE Advanced Pro (Cat 17+). These increased channel bandwidths are achieved using **Carrier Aggregation** (**CA**) where multiple **Component Carriers** (**CCs**) are aggregated together. In 5G, up to 16 CCs can be aggregated together to achieve the highest UE downlink data throughputs exceeding

20 Gbps. In terms of the 5G devices in the market, higher-end 5G devices such as routers and laptops typically support 4 X 4 MIMO and up to 5 CCs 5G in addition to LTE Advanced Pro (Cat 17+) to enable peak DL data rates of around 10 Gbps. Lower-end 5G devices typically support 4 X 4 MIMO with only 2 CCs and with LTE Cat 17+ to enable peak DL data rates of around 5 Gbps. All 5G devices normally support both LTE and 5G, as both networks are used together, as we will describe in this chapter. Let us review the 5G frequency spectrum, which impacts the coverage, capacity, and latency of the 5G network.

5G frequency spectrum

5G networks are typically deployed on three different licensed frequency bands, which are the low bands under 1 GHz, the mid band between 1 GHz and 6 GHz, and the high bands between 24 GHz and 40 GHz, with the band effectively setting the potential range/coverage and data throughput. As with any radio technology, the lower the frequency band, the better the range and penetration through obstacles (such as trees and walls). As shown in *Figure 5.3*, the low- and mid-frequency bands (which are termed **Frequency Range 1 (FR1)** or **sub-6 GHz** and are used today with LTE, 3G, and 2G network deployments) have long ranges for wide-area coverage and moderate bandwidth for good data throughputs. The high-frequency bands, typically between 24 GHz and 40 GHz, which are termed **Frequency Range 2 (FR2)**, are the mmWave bands used primarily in 5G and have very limited ranges—typically less than 1 mile—and limited coverage areas (such as stadiums or dense urban areas). The FR2 bands have the highest bandwidths and enable the highest peak data rates of over 20 Gbps with the lowest latency. An important feature of 5G is **Dynamic Spectrum Sharing (DSS)**, which allows LTE and 5G NR to operate simultaneously in the same bands with the network resources being dynamically allocated to each based on demand. This helps network operators better optimize the use of the 5G spectrum:

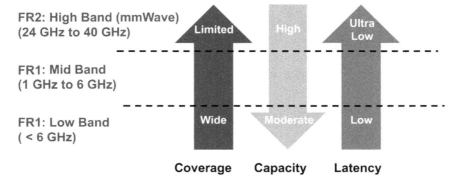

Figure 5.3 – 5G frequency bands

5G network operators have spent billions of dollars to purchase a coveted frequency spectrum, especially in the mid-band and low-band frequencies, and continue to spend billions more on their 5G network infrastructure. Given this significant investment, there is strong carrier support for all 5G services over the next 10+ years. The 5G services market size was valued at USD 47.3 billion globally in 2021 and, according to Grand View Research, is expected to expand at a **compound annual growth rate (CAGR)** of 52.0% from 2022 to 2030 (*Grand View Research, 5G Services Market Size & Share Report, 2022-2030*). One of the key features of 5G SA that benefits both network operators and consumers is an SBA and network slicing where multiple network configurations can be created on top of the common 5G physical infrastructure. Let us now look at 5G network slicing in more detail.

5G network slicing

The 5G core SBA defines a set of **network functions (NFs)** such as the **network slice selection function (NSSF)**, which selects the optimal network functional slice to support services and associated applications, shown in *Figure 5.4*. This network slicing combined with SBA allows for dynamic trade-offs for high data rates and capacity with eMBB, high reliability and low latency (URLLC) with mission-critical control, or high-density (mMTC) services. With the 5G SA core network SBA, a cloud-based, distributed infrastructure is enabled where the various network elements can be deployed at various physical locations relative to the 5G **Base Transceiver Station (BTS)**, as shown in *Figure 5.4*. The 5G BTS or the gNB can be divided into two physical entities called the **Centralized Unit (CU)** and **Distributed Unit (DU)**, which—as with other critical network core elements—can now be moved from the central data center furthest away from the UE to the *edge* data center that is closest to the UE to improve network latency. For example, to achieve the lowest network latency in mission-critical (URLLC) applications, the application server, core user plane, and 5G gNB CU would be deployed in the *edge* data center closest to the UE:

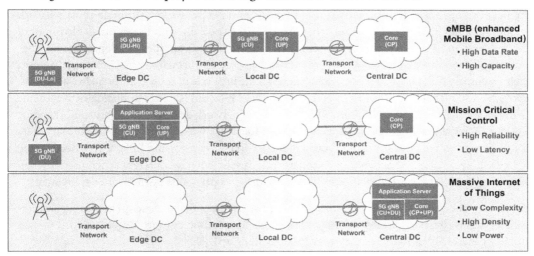

Figure 5.4 – Network slicing for 5G applications

Since 5G NSA architectures don't use a 5G core, network operators don't have the SBA and network slicing features described here. As such, the 5G NSA networks are typically designed to support primarily eMBB services with high peak data rates and capacity, which comprise the most common 5G applications. The full benefits of 5G with low latency (latency on 5G SA is as much as 40% lower than 5G NSA), high peak data rates, and high connection density tailored to the UE application will not be completely realized until networks evolve to 5G SA.

In summary, 5G is a new radio technology that brings unique network architecture features compared to LTE, which enables ultra-reliable, low-latency, and high-bandwidth IoT use cases. Let us now review 5G coexistence with LTE in supporting these use cases before discussing the 5G-specific use cases.

5G and IoT

In our overview of 5G, we described the unique features of the 5G network architecture that can enable new enterprise IoT solutions. In this section, we will look at the 5G technology roadmap and coexistence with LTE in support of new IoT solutions.

Given that 5G is a fairly new technology and the new 5G modules/devices are required to support 4 X 4 MIMO, high bandwidths (high peak data rates), and LTE Advanced (Cat 17+), the cost of these modules/devices is currently around five times higher than standard LTE (Cat 1+) devices/modules. As with any new technology, the cost will decrease over the next few years. Since Qualcomm is the primary chipset supplier for 5G IoT modules and devices, its feature release schedule is a primary driver for the 5G IoT market. The high cost of 5G hardware is a big hurdle for new 5G enterprise IoT solutions where the features of standard LTE and even LTE **low-power wide area** (**LPWA**) (LTE-M or NB-IoT) are sufficient. The primary concern in considering 5G over LTE is network longevity where many network operators have already sunset their 2G and 3G networks. Although most network operators have not announced a sunset schedule for LTE, especially with the initial deployment of 5G NSA reliant on LTE, many enterprise customers such as utilities want the confidence that network operators will support LTE for 10+ years. Most network operators will guarantee LTE network support to their customers for at least 10 years, which supports the life cycle of most enterprise IoT solutions. Moreover, the 3GPP has agreed to allow LTE LPWA (both LTE-M and NB-IoT) technologies to continue evolving as part of the 5G specifications, which means 5G network deployments can include support of legacy LTE LPWA technologies. This is the context for why current LTE LPWA module suppliers claim their modules are *5G-ready*. For your enterprise IoT solution using LPWA, this needs to be validated with your selected network operator, as each carrier will have different schedules for when and how legacy LTE LPWA technologies will be supported as part of its 5G network deployments.

We will review the main 5G eMBB and URLLC enterprise use cases that depend on the unique features of 5G later in this chapter, but, first, we will describe a 5G light technology introduced in 3GPP Release 17 in 2022 called **5G NR Reduced Capacity** (**5G NR RedCap**).

5G NR RedCap

The 5G NR Red Cap release is part of 3GPP Release 17 and introduces a new tier of 5G reduced capability devices. As shown in *Figure 5.5*, 5G NR RedCap is an effort to achieve a compromise between high-end mission-critical URLLC devices and low-end mMTC or massive IoT devices using LPWA. The reduced 5G NR RedCap device capabilities listed here significantly reduce device complexity and cost of 5G devices, thereby covering the IoT market needs currently met by LTE Cat 1 through Cat 4 devices:

- Single transmit-and-receive antenna
- Narrower bandwidths closer to standard LTE (20 MHz in sub-6 GHz)
- Optional support for half-duplex **Frequency Division Duplex (FDD)**
- Lower transmit power

5G Service	Latency	Throughput	Density	Complexity	Battery Life	
5G Mission Critical	LOW	HIGH	LOW	HIGH	LOW	High Performance
5G NR RedCap	MEDIUM	MEDIUM	MEDIUM	MEDIUM	MEDIUM	
5G Massive IoT	HIGH	LOW	HIGH	LOW	LONG	Low Complexity

Figure 5.5 – 5G RedCap fits between 5G massive IoT and mission-critical services

With RedCap 3GPP standardization planned to complete in 2022, the first 5G RedCap modules/devices and supporting 5G RedCap carrier networks will likely not be available until 2024. With 3GPP Release 18 planned for early 2024, further innovation is expected in the IoT segment, with RedCap evolving to potentially replace LTE LPWA technologies. As such, true 5G massive IoT will not happen until 2026 or later, with most networks continuing to serve IoT devices with LTE LPWA technologies. In the context of your enterprise IoT solution, it is important to check with your selected network operator on their schedule for 5G NR RedCap deployment.

Let's review the main 5G use cases that take advantage of the unique high-bandwidth, low-latency features of 5G technology.

5G use cases

The most common 5G use cases focus on the low-latency (URLLC) and broadband (eMBB) capabilities of 5G. Some of these applications, which could be part of an enterprise IoT solution, are as follows:

- Industrial automation (Industry 4.0)
- Remote surgery
- Smart vehicles (autonomous vehicles and **vehicle-to-everything (V2X)**)
- VR/AR

All these applications require very low network latencies (typically less than 20 ms) to enable seamless human-machine interaction. Let's review each use case in more detail in the following subsections.

Industrial automation (Industry 4.0)

Wireless 5G URLLC (as opposed to wired network connections) in industrial automation increases the flexibility and mobility of production lines in terms of where robotic machines can be placed on the factory floor and enables safe human-robot interactions previously only possible with a wired network. 5G technology in a smart factory also enables real-time machine monitoring and remote control, which increases factory operational efficiencies. Moreover, the low latency and high peak data rates of 5G also enable AR for diagnosing and troubleshooting issues on the factory floor.

Remote surgery

Although still in the early stages, remote and robotic-assisted surgeries requiring very low latencies and highly reliable connectivity have already been proven to work over 5G networks, which overcomes geographic limitations for surgical procedures in remote areas. This has benefits to both the healthcare provider and patient in terms of availability and patient recovery time due to higher precision in the operation. There are still issues to overcome before remote surgery is mainstream, but 5G is seen as a viable enabling technology for this to happen.

Smart vehicles (autonomous vehicles and V2X)

Autonomous vehicles and V2X communications are prominent 5G use cases in the context of smart cities where low latency is needed for vehicles to make real-time autonomous driving decisions and act on telemetry with other vehicles and city-wide infrastructure, termed V2X. This has the potential to reduce human error and increase safety (for example, automated braking) with improved traffic efficiency and driver comfort. For example, 5G-enabled vehicles can reroute based on information on traffic congestion from other vehicles or city V2X infrastructure. In the agriculture market, many farming equipment providers are investigating autonomous farming for driverless, non-stop operations. It is clear that 5G will be at the center of future smart vehicle and smart farming use cases.

VR/AR

VR is the computer-generated simulation of a 3D environment with which a user can physically interact using a special headset and body sensors for multiple sensor modalities (visual, auditory, haptic, and olfactory). AR superimposes a computer-generated image over a user's view of the real world in a content-rich composite view. Both VR and AR are ideal 5G NR use cases, requiring both low latency and high data throughput for a seamless, immersive user experience. Both technologies can be part of many IoT solutions, including gaming, remote healthcare (including surgery), smart factory workflows/troubleshooting, and smart city-enhanced resident/visitor experience/safety. Both VR and AR will continue to be primary 5G use cases over the next 10+ years.

Now that we have provided an overview of 5G NR to include the most common IoT use cases, let's discuss some best practices in implementing an enterprise IoT solution with 5G connectivity.

Best practices in 5G solutions

In planning to use 5G NR for your enterprise IoT solution, there are a few best practices in selecting 5G connectivity to avoid unexpected issues later in the solution deployment. First, it is important to identify the requirements for latency, data throughput, and device power consumption (battery life) to determine whether 5G versus LTE (or LTE LPWA) connectivity is needed. In either case, the most important consideration is the selection of your 5G/LTE network carrier.

5G network carriers

As mentioned earlier in this chapter, the most important practice is to validate the specific 5G network features required for your IoT solution with your selected network carrier. Each carrier will support slightly different 5G features (for example, mMTC support) and have different schedules for NSA/SA feature deployments and their evolution from NSA to SA. Moreover, every 5G carrier will support different 5G and LTE frequency bands and have different CA and CC requirements, which need to match with the selected IoT device (such as routers, gateways, or laptops) in your enterprise IoT solution. Most large carriers such as AT&T and Orange have device certification programs that validate device operation on their networks, so it is critical to select 5G/LTE devices that have been validated on your selected carrier's network. If your IoT solution requires global roaming, it is also important to validate the carrier has roaming agreements with carriers in your target countries where the IoT solution could be deployed. Also, the supported technologies (5G, LTE) and frequency bands on the IoT devices must match the supported technologies and frequency bands of the roaming carriers. Once you have selected your 5G network carrier, conducting network trials and pilots of your 5G IoT solution is another best practice that ensures the solution meets expectations. The actual network latency and data throughput will vary significantly based on network congestion, 5G/LTE frequency bands, device capability, and network features, so extensive field testing is important to understand how your IoT solution performs on the selected carrier network. Most carriers allow for IoT solution pilots/trials prior to completing a connectivity agreement.

5G private networks

Private 5G networks can deliver the same features as public 5G networks over a private enterprise-licensed 5G spectrum. Private 5G networks are selected by some industries such as manufacturing or healthcare over concerns about public network security, latency, or bandwidth limitations. The advantages of private networks are a dedicated, licensed spectrum that can be managed by the enterprise for lower latency and better security/privacy. The big disadvantage of private networks is the cost of the network buildout, maintenance, and frequency spectrum, which is typically licensed by one of the major network carriers. There are also regulatory requirements around spectrum availability based on location. In choosing a private 5G network over a public network, it is important to assess whether it really suits the needs of your enterprise IoT solution and justifies the cost and inherent maintenance required over the life cycle of your IoT solution. We will discuss private networks in more detail in *Chapter 10, Looking at the Road Ahead.*

5G IoT devices

Given 5G is a fairly new technology, 5G devices are significantly more expensive than standard LTE and especially LTE LPWA, so it is important to determine whether 5G is actually needed for your IoT solution and the number of devices required to support the solution. Also, many of the current 5G IoT devices only support 5G NSA and not 5G SA, so it is important to ensure the selected devices support the required 5G network architecture and associated features (especially low latency). From a solution cost perspective, if 5G connectivity is required to meet latency and/or data throughput requirements, it is important to find the IoT device with the minimum required 5G NSA/SA functionality in terms of 5G band CC and CA support (for example, 3 CCs versus 5 CCs) and the associated maximum system bandwidth.

Once you select 5G versus LTE based on your enterprise IoT solution requirements, it is important to select your 5G public/private network carrier and validate the required 5G NSA/SA network features and technology roadmap to support the full life cycle of your IoT solution.

Summary

In this chapter, we provided an overview of 5G to include the radio technology, frequency spectrum, and NSA/SA network architectures with network slicing that enable new low-latency, high-bandwidth IoT solutions. This should give you a good understanding of 5G as compared to LTE to select the best technology for your enterprise IoT solution. We then covered the 5G technology roadmap and how LTE and 5G will coexist to enable IoT solutions that should provide a framework for selecting your 5G/LTE network operator. This led to our review of the most common 5G use cases for your enterprise IoT solution to provide insight into potential IoT applications enabled by 5G. We concluded with a review of best practices in selecting 5G in your IoT solution, validating your IoT solution life cycle requirements against the network features supported by your network carrier. Overall, this review should provide the basis for selecting 5G versus LTE and the network operator in your enterprise IoT solution.

In the next chapter, we will cover cellular IoT devices to include the device architectures, types, and network carrier certification requirements. This review should provide a framework for selecting the best IoT device in your cellular enterprise IoT solution using either LTE or 5G technologies.

Reviewing Cellular IoT Devices with Use Cases

In this chapter, we will provide an overview of cellular IoT devices, including a description of the device architecture, device types, use cases, and carrier certifications. We will also cover edge computing, which is an emerging trend in enterprise IoT solutions. As such, we will cover the following topics in this chapter:

- Cellular IoT device architecture
- Cellular IoT device types and use cases
- Cellular IoT device carrier certifications
- Edge computing
- Best practices in cellular IoT devices

In this chapter, you will learn about the components of the IoT device architecture and the various cellular IoT device types, with some information on cellular carrier certifications. We will conclude the chapter with a review of edge computing and some best practices for selecting and using cellular IoT devices in your enterprise IoT solution.

Cellular IoT device architecture

Let us begin our review with an overview of the cellular IoT device architecture shown in *Figure 6.1*:

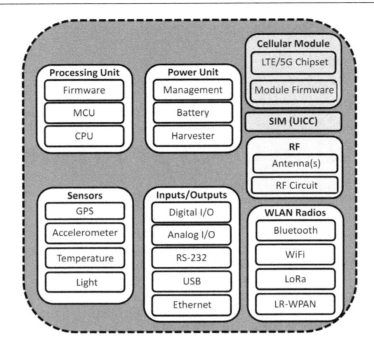

Figure 6.1 – Cellular IoT device architecture

At a high level, a cellular IoT device includes a processing unit that runs the device application, an LTE/5G cellular module with the SIM and RF, a power unit for managing the battery and/or energy harvesting (such as solar or thermoelectric), various sensors, and digital/analog inputs and outputs. It could also integrate WLAN radios such as Bluetooth, Wi-Fi, LoRa, and/or LR-WPAN, which we covered in *Chapter 3, Introducing IoT Wireless Technologies*. Let us examine each of these architecture blocks in more detail.

Processing unit

The **processing unit** is the brain of the device, which manages all the other peripheral architecture blocks on the device and is where the device firmware application runs, on either a **Microcontroller Unit (MCU)** or **Central Processing Unit (CPU)**. A CPU resides in almost all electronic devices, including smartphones, laptops, and even thermostats. They process and execute the instructions in the device application firmware. Depending on the application, a device could have a single CPU core or, more commonly, multiple CPU cores working in parallel for improved performance at the cost of increased power consumption. The CPU performance is primarily defined by its instruction set architecture, which can range from 4-bit to 64-bit architectures. CPU architectures in the 4- to 16-bit range are used in low-cost, constrained IoT devices with simple control loops. CPUs with 32-bit and 64-bit architectures are the most common for IoT applications, as the cost difference between 16-bit and 32-bit is small. An MCU includes a CPU with **Random Access Memory (RAM)**, **Read-Only**

Memory (**ROM**), and **General-Purpose Input/Output** (**GPIO**), which allows the connection of external hardware such as sensors, which are all integrated into a single chip. Due to this computer-on-a-chip integration, in IoT devices, MCUs are a good choice since they provide adequate computing power with low cost, complexity, and power. The most common MCU architectures used in IoT devices are **Advanced RISC Machine** (**ARM**), **Million Instructions Per Second** (**MIPS**), and x86, which is a family of **Complex Instruction Set Computer** (**CISC**) architectures initially developed by Intel based on the 8086 microprocessors.

Most MCUs have either an **Operating System** (**OS**) or **Real-Time Operating System** (**RTOS**), which is an intermediary program between the device firmware application and MCU hardware that manages the hardware and software resources available on the MCU. An RTOS provides a deterministic near real-time response to external events, meaning it can provide a faster response than an OS, which can be an advantage for an IoT device. Many of the constrained IoT devices in enterprise IoT solutions use an MCU running an RTOS. A traditional OS such as Linux is more effective than an RTOS at processing many tasks, but it requires more power, which is not typically a good fit for constrained, low power IoT devices. It is also possible for an MCU to have no OS at all, which is called *bare metal*, which is used when a device has very little memory and/or, due to timing constraints, the application needs to control every hardware resource. For low-power, low-cost IoT devices in your enterprise IoT solution, the selection of the right MCU is critical. In many cellular IoT devices, it is possible to run the device application directly in the cellular module, as most cellular module suppliers have an integrated MCU and have provided an application space on the module for customers to implement their device application. This saves both power and cost in the device component **Bill of Materials** (**BOM**). This brings us to a review of the cellular module.

Cellular module

The **cellular module** in the cellular IoT device, in combination with the SIM and RF, provides connectivity to the LTE and/or 5G carrier networks. A **module** is a single component that bundles the LTE/5G chipset, which is a group of **Integrated Circuits** (**ICs**) that negotiate the lowest-level signaling and authentication with the cellular network. The biggest LTE/5G chipset suppliers are **Qualcomm** and **MediaTek**. Designing a device directly with a chipset requires strong technical expertise and expensive licenses with the chipset suppliers, which can be a significant hurdle. A module overcomes this challenge by bundling an MCU with firmware, memory, a power supply connector, and an antenna port, among other components, on top of the chipset to make it easier for IoT device makers to integrate LTE/5G connectivity. Moreover, modules normally have a simple serial command interface (called **Attention** (**AT**) commands) to the processing unit on the device, making module integration with the device application much easier. When troubleshooting a device connection issue, it is fairly easy to look at the device logs of these AT commands to understand where the issue lies. Modules are much more expensive than chipsets; but, given the ease of integration and lower lifetime volumes, modules are commonly used in IoT devices. Alternately, due to lower cost and economies of scale, nearly all consumer high-volume cellular devices, such as smartphones and tablets from major suppliers such as Samsung and Apple, directly integrate chipsets as opposed to using a module.

There are several LTE/5G module suppliers, such as Quectel and Telit, that provide many LTE/5G module variants from low-end LTE LPWA LTE-M and NB-IoT modules to higher-category LTE and 5G modules. Typically, there are module variants to support the licensed LTE/5G frequency bands in specific geographic regions, such as **Europe Middle East Africa (EMEA)**, **Asia Pacific (APAC)**, **Latin America (LATAM)**, and **North America (NA)**. Many modules also support legacy 2G and 3G cellular technologies for better global coverage as these technologies are still being used in many countries outside of NA. The *global* module variants commonly support 12+ frequency bands and have both 2G and 3G technology fallbacks. These *global* modules are more expensive than the regional variants but offer the best choice for ubiquitous WWAN global coverage. These modules are typically integrated into devices used for global asset tracking use cases such as shipping container tracking. In the case of the low-cost LTE-M and NB-IoT modules, since there is not a single LTE LPWA technology deployed globally, the LPWA module variants typically include both LTE-M and NB-IoT technologies as well as legacy 2G fallback to enable global LPWA coverage. Overall, cellular modules are a key enabler for IoT devices and enterprise IoT use cases and will continue to evolve to support the latest 5G technologies, which are reviewed in *Chapter 5, Validating 5G with IoT*.

Cellular modules typically require both an external SIM and RF circuitry with antennas as part of the integrated IoT device. Let us first look at the SIM component.

SIM

As discussed in *Chapter 4, Leveraging Cellular IoT Technologies*, all LTE and 5G devices require a **SIM** to identify the IoT device on the carrier network and ensure the privacy and security of the data exchanged on the network. This SIM could be a physical plastic SIM in various sizes, an IC that is directly soldered to the device's **Printed Circuit Board (PCB)**, or an **Integrated SIM (iSIM)** that is embedded in the cellular chipset on the module. Most IoT devices use a plastic SIM that is inserted into a SIM holder on the device PCB. This SIM communicates directly with the cellular module to negotiate and authenticate carrier network access. Since a plastic SIM inserted in a SIM holder is susceptible to device vibrations, which could impact the SIM's electrical connection to the cellular module, some IoT device suppliers use a **Machine-to-Machine Form Factor (MFF2)** IC SIM that is soldered to the device PCB. Using an MFF2 SIM has the advantage of being more stable with regard to device vibrations, but it is less flexible in changing SIMs in the device for different network carriers. Using an iSIM offers the advantages of stability and avoids the need for any physical SIM in the IoT device. With regard to network carrier flexibility, any SIM can also be configured as an **Embedded Subscriber Identity Module (eSIM)**, which allows the remote switchover to multiple network carrier profiles. Both the iSIM and eSIM are becoming popular options for all IoT devices, as they save cost and are easier to manage logistically in the manufacture and deployment of IoT devices. We will review the growing iSIM and eSIM trends in *Chapter 10, Looking at the Road Ahead*.

This brings us to the second key external IoT device component that is used with the cellular module for network connectivity, which is the RF block.

RF

The RF block in an IoT device consists of not only the antenna(s) that are tuned to the required frequency bands but also the RF circuitry on the device PCB, which connects the module to the antenna(s). The antenna could be either an internal chip antenna directly mounted on the device PCB or an external antenna with just the antenna connector mounted on the device PCB. The internal chip antenna has a lower gain as compared to an external antenna, but it allows for a self-contained device, eliminating the need for the extra external antenna component. This RF circuitry is designed to correctly match the antenna with the target frequency bands for the best RF transmit power and sensitivity. This circuitry is also designed to mitigate interference from external sources to achieve optimal RF performance of the device. This is important for network carrier device certifications where many network carriers have device RF transmit power and sensitivity requirements in the frequency bands used by the carrier. RF sensitivity is a measure of how sensitive the combined receiver-antenna RF circuity is to the received RF signals. In terms of transmit power, most cellular modules are designed to support 3GPP power class 3 with an output power of 23 dBm, which is 200 milliwatts. With an optimal RF circuitry design, this translates to around 20 dBm or 100 milliwatts of **Total Radiated Power** (**TRP**) at the antenna. Licensed LTE and 5G carrier networks were designed with a minimum expected device RF performance for proper operation of the overall network, which is the primary reason for the carrier requirements for device certifications.

With an understanding of the cellular module and associated RF and SIM components, let us look at the device power unit that manages the power on the IoT device.

Power unit

The **power unit** consists of the circuitry for managing and distributing power to the other components in the IoT device. Each component in the IoT device requires a specific voltage and current level for stable operation, and it is the power unit that manages each of these. In the case of battery-powered devices with a rechargeable battery, the power unit has the circuitry to recharge batteries either from an external power source or an energy harvester component integrated into the device. The energy harvester is a component that converts solar, heat, or vibration energy into electrical energy that is stored in a rechargeable battery. An energy-harvesting device extends the device life cycle and can eliminate the need for replacing batteries periodically. Due to its relatively high-power efficiency compared to other sources, solar energy-harvesting technology is the most commonly used in IoT devices. With low-power devices used in factories with high heat sources, such as steam pipes, heat energy harvesting using thermoelectric components is an emerging IoT use case. Other, less common energy-harvesting sources, such as vibration and kinetic energy, and RF typically do not provide enough energy density for a cellular IoT device. With low-power IoT sensor devices being the focus of many long-life cycle enterprise IoT solutions, the low-power features of the LPWA cellular module and processing unit are critical, especially with energy harvesting.

This brings us to the many sensors that could be integrated into an IoT device, which is the core function of the IoT device in an enterprise IoT solution.

Sensors

There are many sensors that could be integrated into an IoT device, with the most common being a **Global Navigation Satellite System** (**GNSS**) or **Global Positioning System** (**GPS**) receivers for location, accelerometers for movement detection, and temperature, humidity, pressure, proximity, and gas detection sensors. Let us review each of these common sensors.

GNSS/GPS for location

Any enterprise IoT solution, such as asset tracking/monitoring, that requires a location typically uses a GNSS/GPS chipset, which uses RF signals from an array of orbiting satellites to resolve an accurate location. There is a cost and power impact to IoT devices using GNSS/GPS for location which should be considered especially with low-power and low-cost enterprise IoT solutions. In addition to GNSS, many IoT solutions that require locations use a combination of GNSS with **Location-Based Services** (**LBS**) from companies such as Google and Microsoft. These location services use cellular and Wi-Fi data to resolve a device's location where GNSS may not be available. Due to the cost and power consumption of a GNSS chipset, some asset tracking/monitoring IoT solutions use LBS alone to get a location that is not as accurate as GNSS but good enough for the specific IoT solution.

Accelerometers for movement

Accelerometer sensors detect the acceleration of a device and are commonly used to detect the movement or vibration of a device. Accelerometers are typically used with a GNSS receiver to identify when a GNSS-based location is needed since it usually does not make sense to use device power to get a GNSS location when it has not moved. In the context of fleet management use cases, accelerometers are used to monitor driving behavior by detecting vehicle hard-braking and hard-acceleration events.

Temperature

Temperature sensors simply measure the amount of heat in a source and are commonly sensors integrated into IoT devices. Temperature is an important parameter that is monitored in many enterprise IoT applications, including food safety, warehouse inventory management, cold chain logistics, and equipment monitoring, where temperature levels need to be maintained at a certain level for product/asset integrity.

Humidity

Humidity sensors measure the amount of water vapor in the air. They are very common used in **Heating, Ventilation, and Air Conditioning** (**HVAC**) systems and are typically used in combination with temperature sensors to monitor product/asset integrity in IoT asset monitoring and cold chain logistics solutions.

Pressure

Pressure sensors detect changes in gases and liquids and are typically used in remote monitoring use cases for leak detections. They are also used in the energy markets to monitor oil and gas pipeline pressures in the context of utility IoT solutions.

Proximity

A proximity sensor is another common IoT sensor used to detect the presence and range of nearby objects typically using either ultrasonic or infrared beams. For example, a **Pyroelectric Infrared (PIR)** sensor is a common low-cost motion detection sensor used in many IoT applications. There are a wide variety of applications for proximity sensors. Proximity sensors are used to detect the occupancy of parking spots in parking garages and the presence of cargo in a trailer. They can also be used to detect the proximity of a customer to a product in a retail IoT solution to gather marketing data and make special offers of products near the sensor.

Gas detection

Gas sensors measure the presence and quantity of specific gases, usually in the context of air quality in a lab, city, agricultural, or manufacturing environment. The presence of specific gases can have significant impacts on public/work safety and crop yields, which is the impetus for the remote monitoring of gases in smart cities, industrial manufacturing, and agricultural IoT vertical markets.

Data from these sensors and many others can be combined in various combinations to meet the requirements of the enterprise IoT solution use cases. For example, in many fleet management solutions, GNSS/GPS is used for location reporting and accelerometers are used to detect movement, such as to change the device-reporting interval when the fleet vehicle is moving and idle. The combination of sensor data can be used to trigger a reporting event in various IoT use cases. For example, if the monitored asset is moving and goes outside a specific location (called geofencing), it can report unusual activity to the enterprise application. In another fleet management use case, devices with a combination of GNSS/GPS, accelerometer, and ultrasonic proximity sensors are used to track cargo in a trailer and when it is loaded/unloaded at a destination. Clearly, there is a myriad of sensor combinations that could be integrated into an IoT device in the context of an enterprise IoT solution.

In many remote monitoring enterprise IoT solutions, the IoT device has a wired connection to a digital/analog interface, which could be an external sensor, machine, or utility meter. For a broad range of IoT use cases, IoT devices also have a serial or Ethernet interface, which brings us to the input/output block of an IoT device.

Input/output

As we covered in *Chapter 2, Understanding IoT Devices and Architectures*, the most common IoT device types are routers, serial modems, gateways, and remote monitoring devices. Most of these devices support a wired Ethernet, serial, and/or digital/analog interface for many enterprise IoT solutions. Let us review each of these wired interfaces to better understand how they are used in the context of an IoT solution.

Ethernet

Ethernet is the standardized, ubiquitous wired technology for connecting devices in a wired **Local Area Network (LAN)** or **Wide Area Network (WAN)**. It was standardized as IEEE 802.3 in 1983 and can support IP data throughputs of up to 400 Gbps. Ethernet routers or hub devices are common in enterprise **Information Technology (IT)** infrastructure and have at least one Ethernet port, which uses twisted pair cables with a **Registered Jack-45 (RJ-45)** jack. In the context of enterprise IoT, these routers can also have LTE cellular and/or Wi-Fi wireless technologies for WWAN failover solutions that can fall back to these wireless technologies when the wired Ethernet fails. In other IoT use cases where remote Ethernet-based machines such as **Automated Teller Machines (ATMs)** and **Point-of-Sale (POS)** devices require WWAN coverage, an Ethernet router with cellular backhaul is commonly used.

Serial

Serial interfaces on IoT devices use standardized serial protocols such as USB, RS-232, RS-485, and SPI, which were covered in *Chapter 2, Understanding IoT Devices and Architectures*. These serial interfaces, which are used in the IoT serial modem device type, provide a common interface to external peripherals such as **Programmable Logic Controllers (PLCs)** on industrial machines, remote field devices, **Supervisory Control and Data Acquisition (SCADA)** utility field devices, and healthcare devices. The RS-232 serial interface is full duplex and can support data rates up to 3 Mbps, while RS-485 is half duplex and can support data rates up to 40 Mbps. RS-485 has a long range of 4,000 feet, compared to about 50 feet for RS-232. With these advantages, RS-485 is the most common serial interface used in IoT solutions requiring serial interfaces.

Digital/analog

In IoT solutions where the IoT device needs to connect to an external digital or analog sensor, the device needs to have dedicated digital and/or analog interfaces. Although many sensors, such as the ones described earlier in the *Sensors* section, can provide a digital output to the device, there are many low-cost analog sensors like light, sound, and pressure sensors that could be used in an IoT solution. Digital interfaces are more common on IoT devices, as an analog interface requires the device to have an integrated **Analog-to-Digital Converter (ADC)** and/or **Digital-to-Analog Converter (DAC)** component.

In addition to the wired interfaces on an IoT device to connect with external peripheral sensors and machines, many IoT devices also integrate one of the WLAN radio technologies reviewed in *Chapter 3, Introducing IoT Wireless Technologies*, which are Bluetooth, Wi-Fi, LoRA, and LR-WPAN.

WLAN radios

There are many sensors that integrate these standardized wireless technologies to eliminate the need for wired connections and improve the range and flexibility of IoT solution deployments. As discussed in *Chapter 3, Introducing IoT Wireless Technologies*, each WLAN technology has different features in terms of power consumption, range, and scalability. In a gateway IoT device type, these WLAN technologies are directly translated to a cellular WWAN backhaul connection. Bluetooth is commonly used for short-range low-power sensors, which can be powered by coin-cell batteries or energy harvesting. Wi-Fi is commonly integrated into routers for high data-rate WLAN hotspot coverage in conjunction with cellular for WAN backhaul or failover. Wi-Fi is also used in sensors such as security cameras, which require higher data throughput. LoRa is a low-power, long-range technology primarily used for outdoor low data-rate sensor IoT applications. A LoRa gateway device combines this long-range sensor technology with cellular WWAN for backhaul. LR-WPAN is primarily used in indoor, short-range, low data-rate sensor IoT applications such as smart home automation and industrial sensor networks. As with LoRa, an LR-WPAN gateway typically uses cellular WWAN for backhaul.

This concludes our discussion of the cellular IoT device architecture, which leads us to a review of the cellular IoT device types to provide a framework for the device selection in your enterprise IoT solution.

Cellular IoT device types

As described in *Chapter 2, Understanding IoT Devices and Architectures*, in our overview of IoT device types and use cases, the most common device types used in most IoT solutions are serial modems, routers, gateways, remote monitoring devices, and asset trackers. This is true for cellular IoT solutions as well, but to provide more granularity on specific cellular IoT device types that could be integrated into your enterprise IoT solution, we have listed the most common cellular IoT device types in the table shown in *Figure 6.2*:

Alarm / Alarm Panel	Lighting – Indoor	Point of Sale	Tracking – Animal
Camera – Body	Lighting – Street	Remote Control Device	Tracking – Asset
Camera – Dash	Medical Telematics	Remote Monitoring Device	Tracking – People
Camera – Security	Meter – Gas	Router	Tracking – Vehicle
Camera – Traffic	Meter – Parking	Rugged Handheld	Vehicle TCU (Telematics Control Unit)
Camera – Trail	Meter – Power	Sensor – Commercial	Vehicle OBD II (On-Board Diagnostic II)
Computer – In Vehicle	Meter – Water	Sensor – Environmental	Vending Telemetry
Emergency Phone	Modern – Embedded	Sensor – Industrial	Wearable – Fitness
Gateway	Modern – Serial	Smart Home – Appliance	Wearable – Health
Hotspot	mPERS (mobile Personal Emergency Response System)	Smart Home – Thermostat	Wearable – Industrial Safety

Figure 6.2 – Cellular IoT device types

We will now provide a brief definition and explanation of each of these device types, which should provide more insight into selecting the best IoT device(s) for your enterprise IoT solution. Since the general router, gateway, serial modem, and remote monitoring device types were covered earlier, we will not cover these again.

Alarm panel

Both residential and commercial alarm panels are popular cellular IoT use cases, where the cellular technology provides the data and voice primary or backup connection to the alarm central operations center based on alarm events. Given the low data rates required, LTE-M or LTE Cat-1 technologies are commonly used.

Camera

There are many use cases for cellular IoT cameras where remote monitoring without LAN connectivity is important. Cellular cameras include not only security cameras but also body cameras for law enforcement, dash cameras for fleet and ride service solutions, traffic cameras for traffic management in smart cities, and trail cameras for remote monitoring of wildlife. Given the moderate data rates required, LTE Cat-1 to Cat-4 technologies are commonly used in these applications.

Computer – in vehicle

In-vehicle computers are essentially cellular IoT devices with significant computing power that are installed in vehicles and not only monitor the vehicle in terms of engine diagnostics and health but also provide LAN connectivity and fleet management services. These devices typically include vehicle diagnostic interfaces, Wi-Fi and Bluetooth for LAN connectivity, and GNSS for location services. These in-vehicle computers are commonly used in enterprise fleets and law enforcement and use LTE Cat-1 or higher technologies for moderate to high data throughput requirements.

Emergency phone

Emergency phones are typically installed in remote areas such as parks and large campuses where connectivity is challenging. A cellular emergency phone provides cellular voice connectivity without the need for wired infrastructure. Power could be supplied by solar energy harvesting integrated into the emergency phone kiosk. Typically, only **Voice over LTE** (**VoLTE**) is required, which is included in some LTE-M network deployments but is a feature of all LTE Cat-1 and higher technologies.

Hotspot

A hotspot IoT device is simply a cellular device that primarily provides Wi-Fi hotspot coverage for users and could be either externally powered or battery powered. These devices are typically used to provide WLAN coverage where there is none or the WLAN infrastructure is restricted. These are typically LTE Cat-3 or higher devices for WLAN data rates.

Lighting

Lighting control, especially for both indoor building management systems and outdoor smart city solutions, can save significant energy costs by managing unneeded lighting. Given these lighting controls typically need to have their own standalone connectivity, embedding cellular IoT is a logical use case. Given the low data rate of lighting controls, LTE LPWA technologies can be used to save both power and cost.

Medical telematics

There are an increasing number of healthcare IoT use cases where a cellular IoT telematics device makes sense, including remote patient vital sign monitoring, glucose monitoring, blood oxygen monitoring, and medical alerts. With cellular IoT, these devices are not dependent on any WLAN connectivity for reporting telematics data to doctors or hospitals. These are typically low data-rate applications where LTE LPWA or Cat-1 technologies are adequate.

Metering

Remote metering continues to be a strong, high-volume IoT use case where remote cellular IoT devices monitor power, water, and gas without the need for utility personnel to be onsite. This saves significant time and money for utilities globally. In the context of smart cities, remote monitoring of parking, especially in garages, is becoming more prevalent. The metering use case requires very little data and, in most cases, can use LTE LPWA.

Modem – embedded

An embedded modem is a unique device type in that it is a ready-to-integrate cellular modem that can be used in many different IoT applications. As shown in *Figure 6.3*, an embedded modem is an end device that integrates the cellular module, SIM card, and RF connectors/antennas on a single PCB with a connector to make it very easy to integrate into a host IoT device:

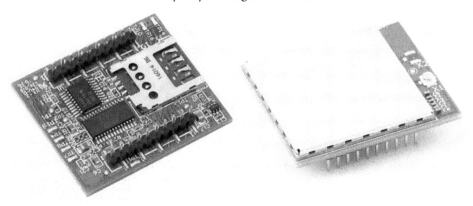

Figure 6.3 – Embedded modem example

Although an embedded cellular modem is more expensive than a cellular module, it is much faster and easier to integrate into a host IoT device, and in many cases, the host device with this integrated embedded modem does not need any further network carrier certifications, as it is typically already certified as an end device. Cellular embedded modems commonly support LTE technologies from LPWA to LTE Cat-4. Embedded modems are great options for initial network testing of your enterprise IoT solution and quickly going to market with a new IoT device.

mPERS

Mobile Personal Emergency Response System (mPERS) devices are wearable devices used in healthcare applications, including senior assisted living and remote patient monitoring, to enable remote alerting to a fall or other medical emergencies. These IoT devices typically support voice and data to a service provider's emergency call center and require low data rates, making LTE LPWA or LTE Cat-1 the most common technology choice for most mPERS devices.

POS

Point of Sale (POS) IoT devices are the devices used at retail locations for processing credit cards and other forms of payment. A cellular POS device removes the dependence of the device on the WLAN network, allowing for remote payment transactions. These devices typically use LTE Cat-1 to Cat-4 technologies.

Remote control device

A remote control IoT device allows for two-way monitoring and control of a server, machine, valve, or actuator, especially in the smart factory, smart city, and agriculture markets. This IoT device type could also encompass indoor and outdoor lighting control applications. These devices typically require low data rates and could use LTE LPWA technologies.

Rugged handheld

Rugged handheld is a broad category of IoT devices, including field ruggedized phones as well as barcode and RFID scanners used in warehouses. The ruggedization usually means the device is more resilient with respect to vibration, dropping, and moisture. These devices primarily run the enterprise IoT application to gather and report field data to the cloud application. The devices can have high data rates and typically use LTE Cat-4 and higher technologies.

Sensors

Sensor devices are another broad category of IoT devices, which could be any IoT device with at least one integrated sensor, such as temperature, humidity, flow, pressure, product levels, or air quality. Nearly all cellular IoT sensor devices are low cost, low power, and battery operated, using LTE LPWA technologies to save cost and power.

Smart home

Smart home devices typically include embedded IoT cellular devices such as thermostats and appliances that provide cellular connectivity for an improved customer experience and service information on home appliances. For example, a smart home IoT device could be a smart refrigerator that can notify the customer when it is time to replace filters or replenish groceries. These devices require low data rates and commonly use LTE LPWA technologies.

Tracking

As mentioned in *Chapter 2, Understanding IoT Devices and Architectures*, asset tracking is currently one of the top cellular IoT use cases based on the number of devices deployed. The tracking device type includes trackers for not only high-value assets but also animals, people, shipments, and vehicles. For the most part, these devices are battery-powered and report periodically and/or based on events such as movement or high temperatures. Given their battery-life constraints, the LTE LPWA technologies are commonly used in these devices.

Vehicle telematics control unit

Most automotive manufacturers integrate cellular in their vehicle **Telematics Control Unit** (**TCU**), which is the brain of the vehicle making the cellular TCU one of the top IoT use cases. The cellular technology embedded in the TCU is used to improve the customer experience with voice calls, vehicle service information, vehicle infotainment, and hotspot coverage.

Vehicle OBD II

All vehicles manufactured after 1996 have an **On-Board Diagnostic II** (**OBD II**) diagnostic port near the steering wheel. This port is used to gather diagnostic information from the vehicle and is also used to power IoT devices plugged into the port. This vehicle OBD II device is commonly used for vehicle tracking and reporting diagnostic data to a cloud enterprise application. This device is also offered by insurance companies to their customers to monitor driving behavior such as hard braking and hard acceleration as part of a **User-Based Insurance** (**UBI**) solution. These devices typically use either LTE LPWA or Cat-1 technologies due to cost constraints.

Vending telemetry

The vending telemetry device type includes IoT devices integrated into vending machines for various consumer products, such as coffee, sodas, snack foods, and even phone charging. This IoT device typically reports to an enterprise cloud application on the vending machine usage, inventory, and health to optimize the vending machine service calls. These devices typically have power but are cost-constrained, so they commonly use the LTE Cat-1 technology.

Wearable

There is a growing number of cellular IoT wearables for health, fitness, and industrial safety. The health and fitness devices are small with several sensors to detect vital signs and fitness parameters, such as heart rate, blood oxygen, movement, and sleep quality. Industrial safety wearables have sensors for noise, air quality, temperature, and alerts. Given that these devices are very power constrained on a small battery, they typically use LTE LPWA technologies.

This overview of cellular IoT device types should provide good insight into the types of cellular IoT devices currently on the market and some context for selecting the best cellular IoT device in your enterprise IoT solution. Since all cellular devices require certification with your selected network carrier, let us now review cellular IoT device carrier certifications.

Cellular IoT device carrier certifications

Each licensed network carrier, such as AT&T, Verizon, T-Mobile, Vodafone, and Orange, has requirements for devices operating on their networks. These requirements are tested as part of a carrier device certification program that validates the device behavior and performance in the frequency bands supported by the selected mobile network operator. Typically, once the device is certified with a specific carrier, it can roam globally on foreign networks that have roaming agreements with the specific carrier. When roaming, the device also needs to support the technology and frequency bands in the roaming country. When devices are deployed globally, there are country-specific regulatory certifications that are typically required in addition to the network carrier certification. For devices that are deployed in several countries as part of a global enterprise solution, it is common to have a single network carrier certification when roaming using the carrier global SIM but multiple country-specific regulatory certifications.

Carrier certifications ensure all devices operating on the network meet minimum requirements to help guarantee the overall network quality for all users. This device certification testing commonly requires the **Federal Communications Commission's (FCC's)** approval, evaluation of the RF performance in the carrier's licensed frequency bands, support for carrier-specific features, and overall device behavior on the network. Some of the device certification tests may be done by the network carrier, but most of the testing is typically done by a third-party lab that is approved by either the **PCS Type Certification Review Board (PTCRB)** or **Global Certification Forum (GCF)** depending on the selected carrier requirement. PTCRB and GCF are forums organized by cellular operators to provide a framework for cellular device certifications. PTCRB was established in 1997 as the certification forum by select North American cellular operators, and GCF was founded in 1999 as the certification forum by mobile network operators, device manufacturers, and test laboratories for certification testing primarily outside of North America. Using PTCRB- or GCF-approved laboratories and certification programs helps facilitate the global interoperability of cellular devices and networks.

Both cellular modules and devices are commonly certified by network operators. When an IoT device uses a cellular module that has already been certified by a network carrier, the certification of the integrated end device using that module is usually much cheaper and easier for the device manufacturer. In this case, the device certification can use much of the testing already completed on the certified module. As such, it is a best practice for IoT device manufacturers to use certified modules as opposed to chipsets in their integrated end devices.

Now that we have described the framework for cellular IoT device certifications, it is important to check with your selected cellular network operator on their specific device certification requirements and whether a device selected for your enterprise IoT solution has been certified with the selected carrier. To conclude our discussion of cellular IoT devices, we will now review edge computing, which is a growing trend in enterprise IoT solutions.

Edge computing

In the context of an enterprise IoT solution, edge computing is where the IoT data collected by the device is analyzed/processed either directly by the device or at computing nodes physically closer to the device. As shown in *Figure 6.4*, any IoT network can be visualized as the IoT end device being the *edge*, with the network/computing nodes between the device and the cloud servers being the *fog* compute nodes where IoT data processing can reduce latency and network load in the overall IoT solution. Compared to the traditional IoT solution architecture, where the IoT device sends all data directly to a centralized cloud server for processing, edge and fog computing can enable more efficient data processing with lower latency for more real-time IoT applications.

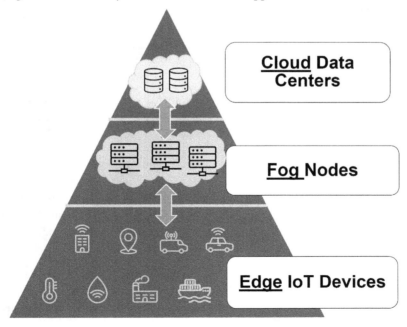

Figure 6.4 – IoT edge, fog, and cloud architecture

Edge computing also enables more robust and scalable enterprise IoT solutions by distributing the IoT processing load to edge and fog nodes and being more resilient to single points of failure in the IoT network, such as the data centers in the cloud or intermittent network connectivity.

Edge computing with IoT devices is typically implemented on devices with significant computing and memory resources, such as a high-end router or gateway, which have a higher cost. With constrained IoT devices with limited compute power in an IoT solution, edge computing is better realized in the "fog" nodes, which could be gateway devices with more processing and storage capabilities. Even with IoT devices with low compute power, there are some lightweight client applications that can run on the device for basic machine learning of the sensor data collected by the device. Using these lightweight machine learning applications, edge IoT devices can detect anomalies in the *normal* sensor data and send notifications to the enterprise cloud application accordingly.

Most LTE/5G network carriers are also enabling **Multi-Access Edge Computing** (**MEC**) for their customers to enable the resilient, low latency, and more efficient data processing of edge computing. As shown in *Figure 6.5*, the MEC servers are between the cellular **Radio Access Network** (**RAN**) and the core network and internet, which connects to the cloud application server:

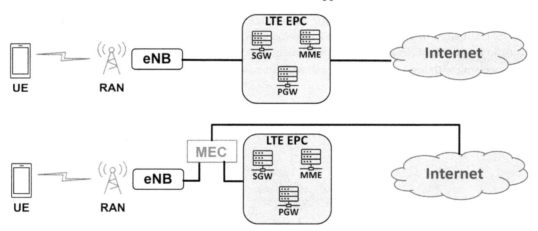

Figure 6.5 – LTE/5G network architecture comparison with MEC

The MEC can be used to host customer applications and make the deployment of enterprise IoT solutions requiring edge computing much easier. The MEC technology benefits IoT customers by offloading the IoT data storage/processing and reducing network latency, but it also benefits network carriers by reducing the network load.

With the evolution of LTE to 5G and the growing number of low-latency IoT applications, especially in the healthcare and smart factory industrial markets, edge computing will continue to be an important part of many enterprise IoT solutions. We will cover edge computing and the closely associated machine learning IoT trends in more detail in *Chapter 10, Looking at the Road Ahead.*

With an overview of cellular IoT devices covering the device architecture, device types, device carrier certifications, and edge computing, let us now review some of the best practices for cellular IoT devices in your enterprise IoT solution.

Best practices in cellular IoT devices

Whether you are selecting an off-the-shelf IoT device or designing your own custom device as part of an enterprise IoT solution, there are a few best practices for ensuring the successful integration of the IoT device in your solution. The first consideration is whether to buy or build your cellular IoT device.

Buying versus building

Designing and building a custom IoT device is a challenging endeavor requiring significant technical resources, time, and money. A typical custom device development takes at least a year and requires additional field testing and pilots to identify hardware and firmware issues for follow-on design iterations prior to large-scale deployments in production. With a custom device, you are also responsible for device support over the full life cycle of the IoT solution. If building a custom IoT device is warranted in your enterprise IoT solution, it is a best practice to use either an embedded modem or cellular module, as described earlier, to make cellular integration and device certification easier. Given the potential pitfalls of building a custom device, it is best to find an off-the-shelf IoT device that can be configured to meet your IoT solution requirements. Many IoT device manufacturers allow their customers to build custom applications on the device.

Once you have selected the IoT device for your IoT solution, the next step is to review any device certification requirements of your selected cellular carrier.

Carrier certification

As discussed earlier, every LTE/5G network carrier has specific requirements for IoT device certifications. Most carriers provide a list of devices that are already certified on their network and will usually help find the best device for your enterprise IoT solution. If your selected IoT device is not certified with the carrier, the device manufacturer will need to complete certification with the selected carrier. If your selected IoT device is not certified and it does not use a cellular module already certified with the carrier, the device certification could be costly and time-consuming. In summary, it is critical to understand the carrier device certification requirements prior to selecting the IoT device(s) in your IoT solution.

Once you have selected the IoT device and validated carrier certification requirements, the next step is to plan device management in your enterprise IoT solution.

Device management

Device management is the mechanism in your enterprise IoT solution to provision, monitor, and maintain the deployed IoT devices over the full IoT solution life cycle. This includes device health monitoring and recovery as well as device remote firmware and configuration updates to fix issues in the field. In many enterprise IoT solutions, the focus is on the device data and cloud application with device management being an afterthought until a device issue is found. It is a best practice to make device management an integral part of your enterprise IoT solution from the beginning. We will discuss device management in more detail in *Chapter 9, Managing the Cellular IoT Solution Life Cycle*.

Summary

In this chapter, we have covered the primary components of the cellular IoT device architecture, the primary cellular IoT device types, an overview of the carrier device certifications, edge computing, and best practices for using cellular IoT devices in your enterprise IoT solution. This should provide a good foundation for selecting, managing, and using cellular IoT devices in your solution.

In the next few chapters, we will cover cellular IoT solution architectures, privacy and security in your IoT solution, and IoT solution case studies to provide real-world examples of enterprise IoT solutions, and an in-depth review of the cellular IoT solution life cycle. We will conclude in *Chapter 10, Looking at the Road Ahead*, with a review of the IoT technology trends and new business models enabled by IoT.

7

Securing the Internet of Things

"Special thanks to Senthil Ramakrishnan for envisioning this chapter and providing much of the content"

– Dennis McCain

As the **Internet of Things (IoT)** continues to grow with an expected 30 billion connected devices by 2025 (source: *Business Insider, IoT Analytics, Gartner, Intel, Statista*), the security of IoT devices and the data they send and receive has become an increasingly critical challenge. As part of an enterprise IoT solution, these devices need to be identified, authenticated, and allowed secure access to enterprise IoT platforms and applications. The IoT data from these devices needs to be sent over secure connectivity channels to ensure data integrity and privacy across the end-to-end IoT solution. As discussed in *Chapter 1, Transforming to an IoT Business*, there is a wide variety of enterprise IoT solutions, including use cases such as remote power sensors, connected cars, and new use cases triggering new security challenges. The security challenges faced by these devices and the associated solution architectures are extensive, and the key to the success of IoT is solving these security challenges and protecting the billions of connected IoT devices. These security challenges cannot be solved using traditional network security solutions and require the development of new solutions that work across the entire IoT ecosystem, which is the focus of this chapter. The lack of standards within the IoT space further complicates the required security solutions, since each deployment is unique as part of a custom enterprise IoT solution.

In this chapter, we will provide an overview of the IoT security ecosystem and the associated security challenges. We will then present an IoT security framework from the device layer to the enterprise IoT application layer. We will conclude with a review of best practices for device security. As such, we will cover the following topics:

- An overview of the IoT security ecosystem
- IoT security challenges
- A proposed IoT security framework
- The best practices for IoT device security

Let us start with an overview of the IoT security ecosystem.

An overview of the IoT security ecosystem

In current IoT solutions, there is not a single, end-to-end security solution. Each point in the ecosystem shown in *Figure 7.1* provides some security solutions that are not connected to other solutions deployed in other points within the ecosystem. This distributed and uncoupled approach is not capable of providing a secure end-to-end system but instead allows points of compromise. In such an insecure system, the critical aspect is the integrity of the data that is being sent and received. Conventional IT data security products such as firewalls and **Intrusion Detection Systems/Intrusion Prevention Systems (IDSs/IPSs)** do not typically transfer over to the IoT space. While these products may play a role in IoT security, the use of newer protocols and varied types of IoT devices requires the development of new solutions. The solutions being deployed will need to account for varied device types and connectivity models.

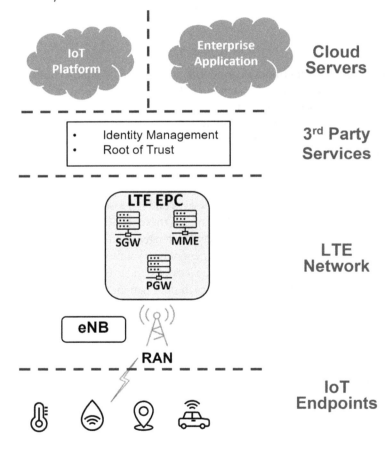

Figure 7.1 – IoT ecosystem

Some of the main IoT security concerns in the IoT ecosystem shown in *Figure 7.1* are as follows:

- Identity and access management
- Data privacy
- Threat management
- Physical device security
- **FOTA/SOTA (Firmware over the Air/Software over the Air)**
- Device management
- Virus/malware
- Internet-based threats

These security concerns will be addressed in the proposed IoT security framework covered later in this chapter. The biggest challenge for enterprise IoT solution deployments from a security perspective is the custom and complex nature of the IoT ecosystem. There is a myriad of IoT device hardware and software combinations, multiple types of operating systems, and many different device applications. It is a significant challenge to secure all of these, but we will describe a security framework and best practices to address this challenge later in this chapter. First, let us review the security challenges in deploying an enterprise IoT solution.

IoT security challenges

Because of the complexity of developing an end-to-end IoT solution, developers and IoT solution/ system integrators typically use readily available IoT components to create their enterprise IoT solutions. Although this is easier, it tends to mean less focus on the integration points for these disparate IoT solution components and closing the security gaps that may exist. As such, security flaws and vulnerabilities go hidden and become part of the overall IoT solution. Once the enterprise IoT solution is deployed, the security risks and vulnerable attack surfaces are exposed.

In the context of an end-to-end enterprise IoT solution, there are several security attack surfaces, including the device and interfaces between the IoT solution architecture components. In *Chapter 6, Reviewing Cellular IoT Devices with Use Cases*, we reviewed the generic components of an IoT device. From a security attack perspective, the device and module firmware, physical inputs/outputs, and WLAN radios could all be vulnerable and can be better secured by avoiding insecure default settings, using secure update mechanisms, and not using outdated components. Specific best practices for securing the components of an IoT device will be reviewed in detail later in this chapter when we discuss the best practices for IoT device security. Referring to the generic IoT solution architecture we reviewed in *Chapter 2, Understanding IoT Devices and Architectures*, security attacks can also target the communication protocols between the architecture components of the enterprise IoT solution. Looking at the IoT platform and enterprise application operating in the cloud, security vulnerabilities in these applications can also compromise the enterprise IoT solution.

The overarching challenge for security in IoT solutions is that as the number of IoT devices scales in volume, so does the number of attack surfaces. With this scale, the risk profile is greatly increased, and security is reduced to the level of integrity and protection provided by the least secure device, which leads to many issues for the enterprise **Information Technology** (**IT**) organizations typically responsible for the enterprise IoT solution. From an IT perspective, some of the challenges specific to IoT devices include the following:

- Visibility and management of IoT devices in a network
- Lack of embedded security on IoT devices, which is difficult to patch
- Managing the huge amount of potentially sensitive data generated from IoT devices
- Managing a diverse set of IoT devices with different features across disparate enterprise teams
- Lack of testing and security validation of IoT devices in a network

One of the biggest challenges is the fact that most IoT devices are not updated or do not have any security patches available to them. Let us now look at the potential threats to IoT devices and solutions.

Threats to IoT devices and solutions

Enterprise IoT solutions create a convergence of the standard IT security threats with **Operational Technology** (**OT**) security threats introduced by IoT, including malware, **Distributed Denial of Service** (**DDoS**) attacks, man-in-the-middle attacks, and spoofing. With the large amount of data generated in an enterprise IoT solution, malware can be more easily hidden, and compromised IoT devices can act together as a swarm to create a significant security threat to an organization. Some common IoT device threats include the following:

- IT and OT convergence
- Ransomware
- Botnets
- AI-based attacks

Let us first look at the convergence of an organization's IT and OT infrastructures with IoT.

IT and OT convergence

As discussed in *Chapter 1, Transforming to an IoT Business*, IoT devices have become common in enterprise *OT* solutions to improve inventory management and factory operational efficiencies. Traditionally, OT and IT networks were separate enterprise systems firewalled from the public internet. With the advent of IoT devices/solutions, these networks have converged, which creates security threats on both networks. This paradigm shift requires an OT/IT holistic approach to security, which we will address later in this chapter.

This brings us to the threat of ransomware.

Ransomware

Ransomware is a malware variant that locks files or devices until a ransom is paid to a bad actor. Since most IoT devices don't have any files stored on them, the cyber-criminal may attempt to simply lock the IoT device; however, this can usually be resolved by resetting the device and/or installing a firmware security patch. Botnets are another well-known security threat for devices.

Botnets

Botnets are IoT devices infected with malware that are controlled by cybercrime groups to carry out DDoS attacks or covert spying, using IoT devices with a camera or microphone. Botnets with IoT devices can be difficult to counter given the high number of IoT devices in a typical enterprise IoT solution deployment. The advent of **Artificial Intelligence** (**AI**) technologies has also created new IoT device threat vectors.

AI-based attacks

While IoT devices are increasingly using AI for edge processing, cyber-criminals are also starting to use AI systems available on the dark web to perform repetitive tasks to execute smart IoT attacks such as DDoS, even mimicking normal IoT data traffic to avoid detection. AI-based attacks are a growing threat in IoT solutions and are the impetus for a holistic IoT security framework, which we will review now.

With all these device security threats and security challenges when implementing enterprise IoT solutions, what is the best way to enable end-to-end security in your IoT solution? This brings us to a review of a practical IoT solution security framework that takes a multi-layered approach to IoT security.

Proposed IoT security framework

The need for an end-to-end security solution that is adaptable to the various IoT connectivity models and varied IoT device types is an essential requirement in any enterprise IoT solution. Security needs to be deployed at various points in the ecosystem, and these security solutions need to be coupled to provide multiple layers of security. The best approach is a multi-layered security framework that covers all points in the IoT ecosystem, as shown in *Figure 7.2*.

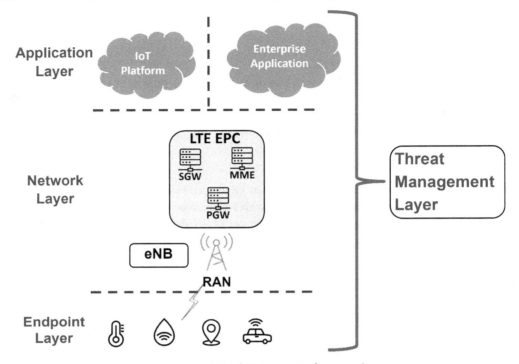

Figure 7.2 – IoT solution security framework

In reviewing this multi-layer security approach, let us start with the device endpoint layer.

Endpoint layer

The physical security of the IoT device is a critical component of any IoT solution. Ensuring that devices are not compromised can help ensure that the data from them is not corrupted. There are various device-side security solutions available today. Some examples of these are as follows:

- **Root of trust**: Secure bootloaders allow the establishment of a base root of trust that the device can leverage. An example is using **Advanced RISC Machine (ARM)** Trust zone.

- **Device identity**: Establishing the endpoint device's identity is key to allowing it to access the enterprise IoT platform and application. Examples would be using a SIM for network access and security tokens/keys/certificates for all IoT devices.

- **Trusted Execution Environments (TEE)**: TEE is an isolated processing environment on the device that helps keep mission-critical components secure and enables a security key requirement to separate secure and non-secure components.

- **Device management**: Having the ability to update and manage the firmware on IoT devices is a key part of maintaining IoT device security. This insures against new vulnerabilities that are identified.

The applications that are running on the device must also be kept secure. These applications ultimately determine the behavior of the IoT devices, and it is critical to lock these down to prevent compromise. Data on IoT devices can be protected using encryption and other methods. For example, the IoT device could use a HW security key to hash the data in device storage and sign any applications running on the devices using this HW security key. Once the IoT data is sent from the device to the enterprise cloud application, the next layer of security is to add access management where the ability to access this data is controlled using a strict policy of identity and authentication. This can be done using secure tokens and certificates.

In an enterprise IoT solution, the endpoint IoT devices do not function in isolation and must work in conjunction with the other layers. This tight coupling is what makes the multi-layered solution secure. This brings us to the network layer in our security framework review.

Network layer

As discussed in *Chapter 4, Leveraging Cellular IoT Technologies*, in a cellular enterprise IoT solution, the IoT device data is sent wirelessly via LTE or 5G network connectivity. The key to security in a network layer is the isolation of IoT traffic from the open internet and the ability to securely transfer data to an enterprise IoT platform and application. This allows a highly secure connection from the IoT endpoint to the enterprise application, keeping data private and isolated from threats throughout the IoT ecosystem.

Cellular networks have inherent security/privacy features from the endpoint to the enterprise cloud application, which we will cover later in this section. Many cellular network operators offer additional network privacy/security options such as **Virtual Private Network** (**VPN**) connectivity between the device endpoint and enterprise IoT cloud application, as shown in *Figure 7.3*.

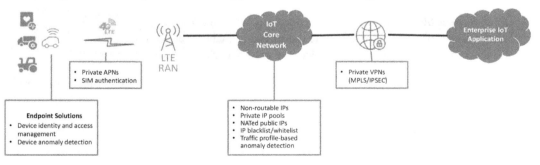

Figure 7.3 – Cellular end-to-end security options

Starting with the IoT device endpoint, some cellular network carriers offer endpoint security solutions with device identity/access management and anomaly detection that are installed directly on a device. In connecting to a cellular **Radio Access Network (RAN)**, network carriers can also provide private **Access Point Names (APNs)** for IoT where each enterprise IoT solution is assigned a unique APN. As discussed in *Chapter 4, Leveraging Cellular IoT Technologies*, an APN is used to enable a connection from the device to the cellular LTE or 5G network. Consumer-type devices such as mobile handsets use shared public APNs. Using a unique private APN for each IoT customer allows the network carrier to provide a separate secure connection from the device to the network. The **Subscriber Identity Module (SIM)**, also covered in *Chapter 4, Leveraging Cellular IoT Technologies*, provides additional security in cellular network connectivity by requiring bi-directional authentication for network access. The device uses the SIM to request network access and resources, and if authenticated, the IoT device is allowed to establish a data/voice/SMS channel to send and receive data. The authentication is bidirectional and allows for a type of mutual authentication between the device and the cellular network. The SIM and the private APN are coupled within cellular systems, ensuring that only approved devices can use a specific APN.

There are some security features in the cellular core network that could be offered by network operators to isolate IoT traffic from the open internet and securely transfer data to an enterprise application. These include the following:

- **Non-routable IP addresses**: Non-routable IPs prevent external sources from communicating with an endpoint device other than the intended IP/port of an enterprise IoT platform/application.

- **Private IP pools**: Compared to public IP addressed, which can be accessed directly over the internet, private IP pools are hidden on the public internet, making it more difficult for an external source to communicate with the endpoint IoT device. Through **Network Address Translation (NAT)**, specific public IP addresses can be mapped to private IP addresses.

- **IP blacklists/whitelists**: This firewall feature allows an enterprise IoT solution to establish approved (whitelisted) and unapproved (blacklisted) endpoints for communication with an IoT device.

- **Peer-to-peer blocking**: Peer-to-peer communication between IoT devices can bypass many network security features and enable the spread of malware and botnet attacks, so blocking direct communications between devices is a good security practice.

Based on this high-level review, we will now take a more detailed look at end-to-end cellular network security, as shown in *Figure 7.4*.

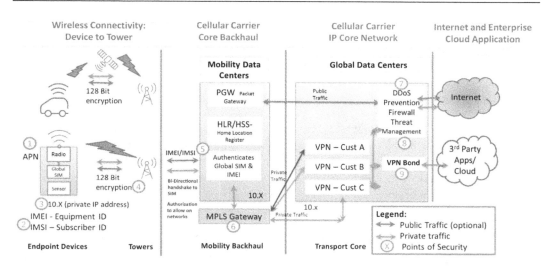

Figure 7.4 – E2E cellular network security

Let us look at each of the inherent security features in *Figure 7.4* in a secure cellular IoT network, starting with a device and ending with data ingestion in an enterprise IoT cloud platform/application:

1. Each device has a private APN that is used to allow it on a private wireless network. This is a private channel from the IoT device to the network.

2. Each IoT device is uniquely identified by its **International Mobile Equipment Identity** (**IMEI**) number, covered in *Chapter 4, Leveraging Cellular IoT Technologies*. Each subscriber is identified by the **International Mobile Subscriber Identity** (**IMSI**) number that is associated with the SIM.

3. The IoT device is assigned a private IP address (for example, 10.x series) that is not advertised on the internet.

4. The IoT wireless data exchanged between the device and the cell tower is encrypted using the 128-bit **Advanced Encryption Standard** (**AES**).

5. The IoT device and subscriber are authenticated, and the traffic is authorized or denied (by the mobility data center in the carrier core backhaul).

6. Private traffic stays segmented through the **Multiprotocol Label Switching** (**MPLS**) gateway flowing to the carrier VPN.

7. The optional network traffic to the internet, if allowed, is protected from a DDoS in the firewall of the carrier IP core network. Any internet traffic uses a NATed public IP that maps the private IP address to a public IP address. IP routing rules and security are in place, and IP/URL blacklists and/or whitelists can be applied to restrict internet-facing data to specific destinations.

8. Multiple layers of security protect the network traffic, which, depending on the network carrier, may include firewalls, network operation centers, security operation centers, and threat management systems.

9. Some network carriers provide a VPN bond between the carrier and enterprise IoT VPNs, so the IoT data traffic stays securely private as it connects to the enterprise IoT application.

Enterprise IoT use cases have introduced several new security challenges and continue to create new ones. Building and deploying an end-to-end security system is becoming an increasingly difficult task, but cellular network providers can provide a multi-layered security framework that addresses many of the IoT solution security concerns presented earlier. Let us now briefly look at the application layer in our security framework review.

Application layer

The application layer is the highest layer of the enterprise IoT solution and is the human interaction component, with unique security concerns for data privacy, integrity, and availability. In addition to unauthorized data access, the application layer is susceptible to security attacks such as DDoS and HTTP floods, which are designed to overwhelm a targeted server. Security implemented at the network layer, such as the private VPNs and firewall IP features described earlier, help prevent application layer security attacks, but adding a more secure multi-factor authentication in the application layer better validates authorized users. It is also a best practice to encrypt IoT data at the application layer to avoid data breaches. Although there is a trade-off between added application layer security and the cost and speed of an application, this is a small price for the privacy/security of potentially highly sensitive IoT data such as medical information.

Security solutions today not only have to counter existing threats but also need to have the ability to identify new threats, using data monitoring to detect intrusions, for example. Traditional IDS/IPS systems need to evolve to include new IoT protocols and device/data behavior for a variety of IoT devices and data types. In the meantime, let us explore some best practices for IoT device security in your enterprise IoT solution.

Best practices for securing IoT devices

IoT device security begins by identifying and applying security practices. Common countermeasures can be used to mitigate threats and reduce a device's attack surface. In general, these countermeasures can be applied to most connected IoT devices as a good starting point to begin securing the device. User access and remote access capabilities along with code validation and event/system logging are at the core of security best practices.

The following table lists some IoT device countermeasures and security control best practices. This list is not comprehensive but can be used to identify core security best practices in your enterprise IoT solution:

Countermeasure	Security practices	Comments
Manage network listening devices	No unnecessary, insecure, or vulnerable network listening services running on the IoT device	• Reduce attack surface by disabling network access to unused services • Services that are exposed on the network are protected by an ACL/firewall – for example, Apache and OpenSSL • Reduce vulnerability exposure by patching/updating applications and libraries exposed on the network
Secure remote access	No insecure/unencrypted remote access to the IoT device	• Communications to/from the device use encryption – for example, SSH as opposed to Telnet
Manage device credentials	Default hardcoded credentials for local access and remote access are unique to each IoT device	• Unique strings provide better protection in cases where the user does not change the password
	For non-unique default credentials, the user must reset the credentials upon login	• If default credentials are not unique, the device must require the user to change the password upon login
Secure sensitive data	No sensitive IoT data sent over a clear-text channel	• Protection must be provided for data in transit – for example, sensitive data should be transmitted using TLS (HTTPS versus HTTP)
	No sensitive data stored in clear text	• Sufficient protection must be provided for sensitive data at rest. For example, private keys and other sensitive customer data should be encrypted using strong, industry-accepted, cryptographic methods, and maintain appropriate file-level access restrictions.
User interface management	IoT device user interface logins can be changed – and the complexity level is required	• Increased security by providing the ability for a user to update their password while also restricting the use of weak passwords
	The user is locked out of the login UI after a specified number of failed attempts	• Reduce exposure to brute-force attacks by enforcing a waiting period between unsuccessful login attempts

Countermeasure	Security practices	Comments
Firmware management	IoT device firmware is remotely updatable and securely transferred	• Firmware updates are critical to patching vulnerabilities and providing new features
	IoT device firmware images are digitally signed and validated	• Authenticating the source of the firmware and the integrity of the firmware code is essential
Open source library management	Open source libraries used in the IoT device firmware/software are accounted for, tracked, and up-to-date to security patch levels	• Reduce attack surface by applying security updates/patches • Tracking all individual applications in use on the device helps ensure that all applications and associated libraries are updated
Access controls	No direct access to the console/JTAG on the IoT device	• Console access should be restricted to authorized personnel only and used only for maintenance and debugging (not exposed to the user)
	Network services are protected by the ACL/firewall on a least privileged (default-deny) basis	• The default "out-of-the-box" mode of operation should deny remote access to the IoT device to help reduce the risk of unauthorized access/DoS
	Remote management of the IoT device is disabled unless explicitly enabled on a per-service basis	• Eliminate unnecessary remote access to reduce the risk of unauthorized access/DoS
	IoT device logging is enabled if the system supports logging	• Event logging allows for the detection/prevention and analysis of unauthorized behavior and access attempts
	Use of Wi-Fi on the IoT device requires minimum security of WPA2 (WPA-AES)	• Eliminate known insecure wireless protocols (WEP, WPA-TKIP, and WPS PIN)
	IoT device connection(s) to external systems are authenticated on both the client and server sides	• An example is communication with management systems; the client should be authenticated as well as the management server

Let us now explore some more specific IoT device security best practices covering primary IoT device components:

- Hardware
- Firmware/software
- Operating system
- Physical interface connectivity
- Network connectivity

These device security best practices help protect an IoT device application from unauthorized access via the various device components, where a bad actor could, for example, take control of the IoT device for malicious purposes, such as disabling a vehicle or creating a large swarm attack using hijacked devices.

We will start with IoT device hardware security.

IoT device hardware security

In this section, we will cover some best practices to ensure the security of IoT device hardware:

- A device's processor system has an immutable secure boot process that is enabled by default. The device's processor system has an immutable trusted root hardware secure boot process.
- The device's processor system has a measured immutable hardware secure boot process.
- The debug interfaces on devices such as JTAG are locked down to only communicate with authorized entities on pre-production devices.
- All the physical input/output interfaces on the device such as USB and RS-232, which are not used in normal operation, are locked down to only communicate with authorized entities.
- The production device test points are disabled or removed when possible.
- When the device security keys are stored outside the processor, the keys should be cryptographically paired to prevent the keys from being used by unauthorized software.

IoT device software security

In this section, we will cover some best practices to ensure the security of IoT device software:

- A device has security mechanisms to block unauthorized software or files from being loaded on it
- The IoT device software images for remote upgrades are digitally signed by an approved signing authority
- The IoT device software images for remote FOTA updates should be encrypted over the air

- The IoT device firmware signing root of trust and crypto keys are stored in tamper-resistant memory

- The IoT device should have mechanisms to prevent software reversions that could be less secure

- To prevent the installation of pre-production software onto production devices, the crypto keys used for signing production software should be unique compared to test or development software

- When development software is used on production IoT devices, the debug functionality should be switched off

- The IoT device should use public/private key equivalents when communicating to an IoT platform/enterprise application over TCP/IP or UDP/IP

- The IoT device crypto keys for software update integrity should be managed securely according to industry security standards such as FIPS 140

- There should be established procedures for how an IoT device is updated either over a network connection or physical interface using a secured input/output port

IoT device OS security

In this section, we will review some best practices to secure the IoT device operating system:

- The IoT device OS should have the most current patches prior to production release

- There should be an ongoing process for validating and updating IoT device OS patches

- All accounts and logins should be disabled or removed from the OS prior to the production release

- All the device OS files and directories should be set to the appropriate access levels

- All password files should only be accessible/writable by the device OS's most privileged account

- All the device OS services not used in operation should be removed from the software image and files

- All command-line interfaces to the OS with the most privileged account access should be removed

- All the device OS kernels and functions should be restricted from being called by external interfaces and unauthorized applications

- The device applications should be operated at the lowest privilege level needed

- All the applicable OS security features are enabled

- The device OS should be separated and secured from the actual application

IoT device connectivity interface security

In this section, we will cover some best practices to ensure the security of the IoT device connectivity interfaces:

- For IoT devices with network interfaces, the unintended packet-forwarding functions should be blocked

- Devices should support the latest application protocols with no known security vulnerabilities, and it should not be possible to downgrade these protocols to a less secure version

- Avoid using insecure and unauthenticated application layer protocols such as **Teletype Network (TELNET)**, **File Transfer Protocol (FTP)**, **Hypertext Transfer Protocol (HTTP)**, and **Simple Mail Transfer Protocol (SMTP)**

- All the unused ports on the device should be blocked or deactivated

- If a device connection requires authentication, the factory or reset password for the connection should be unique to each device and not derived from a number associated with the device, such as its IMEI

- The connection authentication passwords should be changed from the factory or reset password in a normal operation

- If the IoT device uses the MQTT protocol for reporting, it should use **Transport Layer Security (TLS)** with no known security vulnerabilities

- If the IoT device uses the CoAP protocol for reporting, it should use **Datagram Transport Layer Security (DTLS)** with no known security vulnerabilities

- When the TLS or DTLS security protocols are used, they should be validated against the latest security recommendations, such as NIST 800-131A

- After production launch, maintain the latest security protocols for all IoT device connections

IoT device network connectivity security

In this section, we will cover some best practices to ensure the security of the IoT device network connectivity:

- Isolate the traffic from a device to the backend server via point-to-point networks (MPLS VPN) or end-to-end encryption (IPSec or Internet protocol security VPN)

- Avoid direct internet access for the device

- The use of shared APNs should be avoided to allow for IoT solution-specific security policies

- Disable all peer-to-peer communication to prevent direct device access without backend involvement

- Avoid the use of public static type addressing that allows direct internet access from the device

- Use blacklists or whitelists in the firewall to prevent unauthorized access to the device

- Use port-based firewalling to prevent any unauthorized access, which is only possible when using private non-shared APNs

- When possible, disable any device-terminated data access methodologies

- For non-secure protocols such as SMTP, use additional security layers such as encryption of data and/or enable SMS closed user groups

IoT device Wi-Fi security

For IoT devices with integrated Wi-Fi, we will now cover some best practices to ensure the security of Wi-Fi connectivity:

- Enable a firewall within the Wi-Fi chipset

- Enable a **Demilitarized Zone (DMZ)** for the Wi-Fi hotspot so that all internet browsing traffic is isolated from the IoT device

- Isolate clients on the Wi-Fi network from communicating with each other using **Public Secure Packet Forwarding (PSPF)**

- Use network segmentation if supported by the Wi-Fi chipset

- Enable **Wi-Fi Protected Access 2 (WPA2)** encryption with a **Pre-Shared Key (PSK)**

- Disable all web-based admin access

- Disable all default passwords

- Disable services such as Telnet, **Universal Plug and Play (uPnP)**, **Secure Shell (SSH)**, and **Home Network Administration Protocol (HNAP)**

- Use strong password requirements

- Limit the Wi-Fi RF signal to a 5-foot radius

This concludes our review of specific best practices in IoT device security in the context of an enterprise IoT solution. It should be clear that end-to-end IoT solution security is best achieved by using a distributed and multi-layered approach. Let us now summarize our review of the IoT security ecosystem challenges and best practices.

Summary

It should be clear that security is one of the biggest challenges in deploying an enterprise IoT solution. In this chapter, we covered the IoT security ecosystem and its challenges, along with a practical security framework and best practices to address these challenges. These challenges can be addressed by following a multi-layered approach, using the specific IoT device security best practices presented. Here are some of the key takeaways from this chapter:

- Security is a critical part of an enterprise IoT solution and should be a key focus in the planning phase of the solution

- There is no single security solution for an IoT solution, so end-to-end security is optimally achieved by using a distributed and multi-layered approach

- IoT security is constantly evolving, and security solutions need to be adaptable and upgradeable

- Separation and isolation of IoT data can help ensure data integrity and privacy

- Identity and access management are key parts of an end-to-end secure IoT solution

In the next three chapters, we will review two real-world IoT solution case studies, followed by a discussion of the cellular IoT solution life cycle with best practices. Lastly, we will review the new IoT technology trends and business models enabling future enterprise IoT solutions in *Chapter 10, Looking at the Road Ahead*.

8

Implementing an IoT Solution with Case Studies

To crystallize our review of cellular IoT, including the enterprise markets, architectures, network technologies, devices, and security presented thus far, we will now present some real-world enterprise IoT solution case studies. We will start by describing the business cases driving the need for cellular IoT solutions and then cover the architecture and technology decisions leading to the solution's implementation and life cycle management. We will conclude with a review of the lessons learned from these case studies to provide insight into how to create a more robust enterprise IoT solution. For these case studies, we will cover the following topics in this chapter:

- Exploring the IoT solution business cases
- Investigating the device architectures
- Analyzing the network architectures
- Lessons learned

Exploring the IoT solution business cases

We will present two real-world IoT solution case studies that cover business cases in the popular asset management and supply chain logistic markets. The first case study is the **Connected Cooler** solution, while the second case study is the **Smart Label** solution. Both solutions utilize LTE technologies with custom IoT devices using internal/external sensors and location-based services. The Connected Cooler solution uses an IoT device that is powered and feature-rich to support robust cooler monitoring globally, while the Smart Label solution uses a low-cost, battery-powered device that monitors single global shipments. Let us begin by reviewing the business cases and high-level requirements for these two solutions.

Connected Cooler

In the Connected Cooler solution, the high-level business case is to track the location of and monitor beverage coolers manufactured and deployed globally to better manage the supply chain logistics of the coolers and monitor the usage and temperature of the coolers once deployed in retail locations. Data from the connected coolers must support business decisions on where the coolers should be deployed, what products should be sold in the coolers, which coolers have the best **Return on Investment** (**ROI**), and the product integrity in the coolers. Given the business case, the Connected Cooler solution has the following high-level IoT requirements:

- Integration of the IoT device with sensors in the cooler
- Global device connectivity
- Global location and sensor reporting to the enterprise cloud application
- Low-power operation mode for reporting when the cooler is not powered

Let us review each of these requirements in more detail to scope the enterprise IoT solution in terms of connectivity technologies and architecture.

Integrating the IoT device with sensors in the cooler

In this solution, the IoT device will need to be integrated into the coolers with sensors to detect the temperature inside the coolers for product integrity, as well as cooler usage. The easiest way to monitor cooler usage is to detect the number of door openings, which requires a door sensor. To avoid tampering or damage once the cooler has been deployed, the IoT device will need to be covertly installed inside the cooler by the cooler manufacturer. Once installed, the IoT device will gather the temperature and door opening sensor data periodically and report to the enterprise cloud application, which brings us to the requirement for global connectivity.

Global device connectivity

Since the cooler could be manufactured and deployed globally, the integrated IoT device will need to have global connectivity for reporting. This means the wireless connectivity needs to be a public cellular WAN technology such as LTE, 3G, and/or 2G that can work in the countries where the coolers are manufactured and deployed. Other unlicensed WWAN technologies such as LoRa cannot support this global connectivity requirement given their limited global coverage areas. This takes us to the requirement for global location and sensor reporting.

Global location and sensor reporting to the enterprise cloud application

A key requirement for the Connected Cooler solution or any asset management solution is location and sensor reporting. In this solution, the coolers will be deployed at indoor retail locations where the installed IoT device will likely not be able to get a reliable GPS location. As discussed in *Chapter 6, Reviewing Cellular IoT Devices with Use Cases*, for IoT solutions that require both indoor and outdoor locations, **Location-Based Services** (**LBSs**) that use cellular and Wi-Fi data to resolve a device's location, where GPS may not be available, are commonly used. As such, the IoT device in this solution will need to be able to report cellular and Wi-Fi data to enable LBS in the cloud IoT platform. In addition to location, the integrated IoT device will need to connect to external sensors for cooler temperature and door openings. These wired sensors will need to be installed and routed in the cooler with the IoT device. Additional sensors such as cameras could be added as part of the Connected Cooler solution to provide more data on product inventory inside the cooler as well as cooler health, so the IoT device should have the interfaces to support these future requirements. Once the cooler has been manufactured, the device will not be powered until it arrives at the retail location, which brings us to the low-power operation mode requirement.

Low-power operation mode for reporting when the cooler is not powered

Since the cooler will be unpowered when stored and shipped after being manufactured, the IoT device will need to have a battery and low-power mode for reporting in the interim period before it is deployed in the retail store. In this Connected Cooler solution, it could take up to 6 months for the cooler to be deployed, so the device battery and low-power mode should be able to support 6 months of reporting. This low-power mode could be achieved by extending the reporting interval and putting the device in sleep mode until the next report. Moreover, once the cooler has been deployed in the retail store, it could be powered off for various reasons, in which case the IoT device needs to go into a similar low-power battery mode until it is powered again.

These high-level Connected Cooler solution requirements will drive the decisions on both the IoT device and network architecture, which we will cover in the next section. But first, let us review the business case for the second case study regarding the Smart Label solution.

Smart Label

In the Smart Label solution, the business case is to track and monitor shipments globally to better manage supply chain logistics and monitor shipment integrity. As shown in *Figure 8.1*, this IoT solution should be able to identify the shipment location along the supply chain, when it has arrived at its destination, when it has been opened/tampered with, and whether it was subjected to high vibrations/temperatures exceeding thresholds for product integrity:

Figure 8.1 – Smart Label business case

Examples of shipments that would be good candidates to use this Smart Label solution are perishables such as medications or produce and sensitive, high-value assets. The business case for this Smart Label enterprise solution is ensuring product integrity for the end customer and better managing supply chain logistics. Based on this business case, the solution needs to work globally and be relatively small and inexpensive to affix as a label on many shipments. Like a shipping label, the device would ideally be cheap enough to be disposable for each shipment. This drives the following high-level requirements for the Smart Label solution:

- Global connectivity
- Global location and sensor reporting
- Low-power operation with battery
- Low-cost for one-time use

Let us review each of these requirements in more detail to scope the enterprise IoT Smart Label solution.

Global connectivity

Like the Connected Cooler solution described previously, the Smart Label IoT device needs to support global connectivity, as shipments could be shipped anywhere in the world. This means the wireless connectivity needs to be a public cellular WAN technology such as LTE that can work globally; however, since the Smart Label is battery-powered, it also needs to be designed for low-power operation over the life of the label. As we discussed in *Chapter 4, Leveraging Cellular IoT Technologies*, the global LTE LPWA technologies LTE-M and NB-IoT are a good fit for this connectivity as they are both low-cost and low-power technologies. This brings us to the requirement for global location and sensor reporting.

Global location and sensor reporting

Like the Connected Cooler solution, the Smart Label IoT device needs to report location and sensor data to the enterprise cloud application. In terms of location, as discussed in *Chapter 6, Reviewing Cellular IoT Devices with Use Cases*, there is a cost and power impact to using GPS for locations, so in the Smart Label solution, it is best to exclusively use LBS to resolve locations in the cloud IoT platform based on cellular WAN data. Although these locations are not as accurate, the LBS locations are typically *good enough* for shipment tracking. In terms of sensors, the Smart Label device needs to have integrated sensors for temperature, shock, and opening/tampering to validate product integrity during the shipment. Given the need for a low-cost, single label to affix to a shipment, all these sensors need to be internal to the Smart Label device. In terms of cellular connectivity, location, and sensor reporting, the Smart Label needs to be low power to support shipments and storage, which can take several months. Now, let us review how we can enable this low-power operation.

Low-power operation with battery

As discussed in *Chapter 4, Leveraging Cellular IoT Technologies*, there are features of the LTE-M and NB-IoT technologies that can enable low-power IoT use cases. In the case of the Smart Label, the device can use **Power Saving Mode** (**PSM**) with either LTE-M or NB-IoT technologies to stay in low-power mode for up to 413 days, depending on the network carrier's implementation of this feature. Once the device detects movement using an integrated accelerometer sensor indicating the shipment has started, the device can then start reporting at periodic intervals while still using both the LTE LPWA PSM and eDRX features to save power consumption. Depending on the reporting interval and time for shipment, the Smart Label could last for several months on a small battery. In cases where the Smart Label is not able to connect to an LTE-M or NB-IoT network, the device can simply log the location and sensor data for a later report to avoid excessive network searching, which consumes power. To further save both cost and power, the Smart Label can avoid using GPS and Wi-Fi for location and exclusively rely on LBS with cellular WAN data to resolve a rough location for the shipment. Now, let us look at how to reduce the cost of the Smart Label to enable large-scale use on many shipments.

Low-cost for one-time use

Ideally, the Smart Label would be cheap enough to be a one-time-use disposable/recyclable device to enable large-scale adoption on millions of shipments globally. This can be partially achieved by using LTE LPWA technologies, eliminating radios such as GPS and Wi-Fi, which are typically used for location, and using a small integrated battery of less than 200 mAh. Moreover, designing the Smart Label without a mechanical enclosure and using components that are easily mass produced such as printed antennas and even printed polymer batteries enables the economies of scale required to bring the device cost below $10 in large volumes. The combination of low-cost design with low-power operation can enable the mass adoption of the Smart Label IoT solution for a subset of the billions of daily global shipments.

With the business cases and high-level requirements for the Connected Cooler and Smart Label solutions covered, we will now review the device and network architectures for these solutions while leveraging the review of cellular IoT devices in *Chapter 6, Reviewing Cellular IoT Devices with Use Cases*, and the cellular IoT network architectures in *Chapter 2, Understanding IoT Devices and Architectures*.

Investigating the IoT device architectures

For the Connected Cooler and Smart Label solutions, we will review the device architectures using the architectural block diagram (*Figure 6.1*) presented in *Chapter 6, Reviewing Cellular IoT Devices with Use Cases*, and shown again in *Figure 8.2* here:

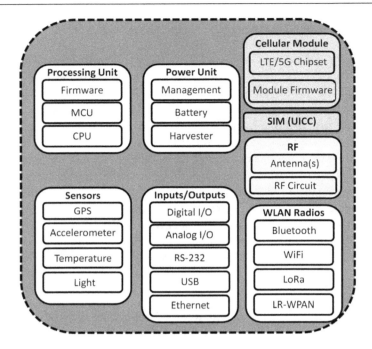

Figure 8.2 – Cellular IoT device architecture

Let us start by reviewing the Connected Cooler device architecture.

Connected Cooler

Based on the business case and high-level requirements presented in the previous section, we can remove some device components shown in *Figure 8.2* to create the more simplified architecture shown in *Figure 8.3*:

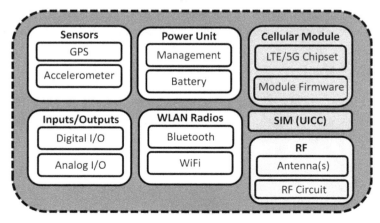

Figure 8.3 – Connected Cooler device architecture

At a high level, the Connected Cooler IoT device needs to support global connectivity and report location and the temp/door sensors integrated into the cooler. Let us review each of these components in more detail.

Cellular module

For global connectivity, the Connected Cooler device needs to support cellular WAN, where the device is manufactured and deployed anywhere in the world. Ideally, the cellular module would support all cellular WAN technologies, including LTE, 3G, and 2G, in multiple frequency bands for the best global coverage; however, the cost of the module/device with this technology support would be quite high. To reduce the cost of the module/device, a best practice is to either select a module that supports a subset of cellular technologies/frequencies, which provides a reasonable compromise on global coverage, or design multiple variants of the IoT device with support for the specific global regions, including North America, Europe, Middle East, and Asia. In this Connected Cooler case study, we decided to create two variants of the Connected Cooler device with a variant for North America and another variant for countries outside of North America. The North American variant would support LTE Category 1 in bands 2, 4, 5, and 12 with 3G fallback in bands 2 and 5. The **Rest of World** (**RoW**) variant would support 3G in bands 1, 2, 5, and 8 with 2G fallback in 4 bands. Creating multiple variants reduces the cost of the module/device given the limited technology/frequency support; however, it also means the North American variant will not necessarily work outside of North America and vice versa for the RoW variant. In terms of the cellular WAN technology selection, the data throughput requirements of the Connected Cooler solution are low, which means LTE LPWA or Cat 1 technologies would suffice. LTE Cat 1 was selected since it has better global coverage compared to LPWA (LTE-M and NB-IoT), as discussed in *Chapter 4, Leveraging Cellular IoT Technologies*. To further reduce the cost of the Connected Cooler device, we also decided to implement the device firmware application on the cellular module using the application space provided by the module supplier. This eliminated the need for an additional processing component such as an MCU or CPU. Now, let us look at the SIM component.

SIM (UICC)

For this IoT solution, the common 3FF form factor plastic SIM was selected for use in the Connected Cooler IoT device. Given the vibration of the device during installation, shipping, and deployment, using a soldered MFF2 form factor SIM would have been a better choice, but a 3FF plastic SIM was cheaper and allowed more flexibility in switching cellular network carriers or upgrading SIMs if needed in the future. Now, let us look at the RF component.

RF

Since this IoT solution will be using an LTE Cat 1 module, it will have two fixed RF antennas on the Connected Cooler IoT device. These two antennas allow for the antenna diversity required for all LTE Cat 1 devices and allow for better RF radiated and sensitivity performance compared to single-antenna devices. The antennas and RF circuitry that connect to the cellular module described previously are tuned to the frequency bands required by the Connected Cooler solution. This takes us to the sensors on the Connected Cooler IoT device.

Sensors

As described in the high-level requirements, the Connected Cooler device primarily relies on external temperature and door opening sensors, which translates into digital and analog inputs for the Connected Cooler device, as we will cover in the next section. In terms of internal sensors, the device needs GPS for accurate locations when there is GPS coverage and an accelerometer to detect unauthorized cooler movement after it has been installed in the retail store. Since the Connected Cooler is deployed indoors, GPS coverage is not available in most deployments, which means the device also needs to support LBS with Wi-Fi, which we will cover when we look at the WLAN radio component. Now, let us look at the input/output component, where the Connected Cooler device interfaces with external sensors.

Input/outputs

In the Connected Cooler solution, there are two external sensors that the IoT device needs to poll periodically. These are the temperature and door-opening sensors. In this solution, as well as most IoT solutions reporting temperature, the temperature sensor is analog, meaning the device needs to have an integrated **Analog-to-Digital Converter** (ADC) to convert the analog temp value into a digital value that is averaged each hour and reported to the enterprise cloud application. The door opening sensor is a digital sensor, meaning it reports a **1** when the cooler door is opened and a **0** when the cooler door is closed. The Connected Cooler device accumulates the number of door openings and reports both the average temperature and door openings in hourly reports. To support future analog/digital sensors such as proximity, level, and infrared, it is a best practice to include some additional input/outputs. So, in this solution, we specified two analog and seven digital inputs. This brings us to the WLAN radio component.

WLAN radios

One of the primary requirements of the Connected Cooler solution is location reporting where GPS may not be available, so the device needs Wi-Fi to support LBS for location resolution in the enterprise IoT platform. With Wi-Fi, the Connected Cooler device can report the unique **Media Access Control** (MAC) address and signal strength of Wi-Fi **Access Points** (APs) in the vicinity of the Connected Cooler. Combining this Wi-Fi data with cellular WAN data in the LBS used by the enterprise IoT platform provides a fairly accurate location of the Connected Cooler globally. To help locate the Connected Cooler in a retail store, we also decided to integrate a Bluetooth radio for beaconing, which can be used by enterprise applications on a smartphone to locate the cooler in the store for product promotions. Now, let us look at the Connected Cooler power unit.

Power unit

Once the Connected Cooler has been deployed in the retail store, it is expected to be powered continuously. However, after manufacturing during shipment and storage, the cooler will be unpowered, which means the Connected Cooler device needs an integrated battery to report locations along the supply chain. When unpowered, the Connected Cooler device will need to be in low-power mode, where the reporting interval is extended from daily to weekly or more to conserve battery power.

Once the device has been powered, it can continue to report daily or more frequently. This means the Connected Cooler device needs an integrated battery and charging circuitry to support unpowered periods during manufacturing, shipment, storage, and deployment. In the most aggressive power saving mode, to allow unpowered reporting for several months, we decided to turn off all cellular and WLAN radios, only waking them up every 2 weeks to report their location and receive any queued cellular WAN messages.

This concludes our review of the Connected Cooler device architecture. Now, we will explore the device architecture of the Smart Label, which has similar location and sensor reporting requirements with a very low-power, low-cost device architecture.

Smart Label

Based on the business case and high-level requirements presented in the previous section, we can streamline the device architecture in *Figure 8.2* to the simplified architecture shown in *Figure 8.4*:

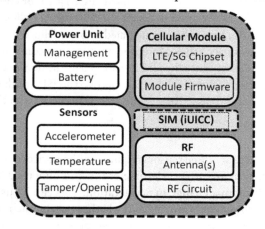

Figure 8.4 – Smart Label device architecture

At a high level, the Smart Label IoT device needs to support global connectivity and report location with temperature, shock, and tamper/opening sensor data for shipments. Let us review each of these components in more detail, starting with the cellular module.

Cellular module

Unlike the Connected Cooler device, the Smart Label is exclusively battery-powered, so it needs the lowest power cellular WAN module technology to save both power and cost. With its inherent power-saving features, LTE LPWA (LTE-M and NB-IoT) is the best technology choice from a power and cost perspective while also enabling global coverage. Since neither the LTE-M nor NB-IoT technology is deployed globally, the Smart Label module needs to support both technologies and multiple LTE frequency bands for global operation. Even with the inherent power-saving features of LTE LPWA, the module design also needs to have a low operating current in its low-power mode, which should be evaluated in selecting the best LPWA dual-mode module. Now, let us look at the RF component in the Smart Label.

RF

Since the Smart Label device is using a low-power LTE LPWA cellular module, it only needs to support a single RF antenna. To reduce the cost of the device in large production volumes, this RF antenna was printed inside the label along with the battery, as we will describe shortly. The antenna and RF circuit connected to the cellular module were tuned to support the multiple LTE LPWA frequency bands for global operation. This takes us to the SIM component.

SIM (iUICC)

While physical SIMs, both removable and soldered, are the most common SIMs used in IoT devices, in the case of the Smart Label, the **Printed Circuit Board** (**PCB**) area for components is limited, so using an integrated SIM or iSIM, as reviewed in *Chapter 4, Leveraging Cellular IoT Technologies*, is a good choice. An iSIM has the advantage of being integrated into the cellular module chipset, removing a physical component from the Smart Label. Moreover, an iSIM also consumes less power, which is important in the Smart Label solution. Since the SIM is managed by the network carriers, it is important to determine whether the selected network carrier has approved an iSIM with the chosen LTE LPWA module/chipset in the Smart Label device. Now, let us look at the sensors in the Smart Label.

Sensors

In the Smart Label solution, there are three sensors required in the device to detect movement/shock, tampering/opening, and temperature. An accelerometer is commonly used in IoT devices to sense movement and shock, resulting in a change in acceleration. This can be used in the Smart Label solution to identify when the shipment is moving and whether the shipment is exposed to excessive shocks, such as a drop. To detect shipment opening and tampering, an electrical trace can be routed at the edge of the Smart Label, which would be cut if opened or tampered with. In cutting this trace, the device can detect an "open" circuit and report accordingly. Finally, the temperature sensor is used to detect whether the shipment is exposed to excessive temperature ranges, which would impact shipment integrity. These sensors consume very little power and can be polled at various intervals to optimize the sensor readings based on the shipment environment. This brings us to the power unit, which manages the low-power operation of the Smart Label.

Power unit

Given the constraints of the printed-battery technology chosen for the Smart Label to enable low cost and mass production, the integrated battery capacity is less than 200 mAh, which means power management is critical. The power unit must manage this limited battery capacity to allow the Smart Label to last for at least 6 months of shipment and storage. This means the power unit must keep the device in a low-power sleep mode, only waking on specific events such as movement or opening/

tampering. For example, when the shipment is in storage and not moving, the Smart Label device does not need to report and can be kept in sleep mode, waiting for the movement to start reporting.

Unlike the Connected Cooler device, to save both cost and power, the Smart Label does not include a GPS or Wi-Fi radio component, which means the Smart Label location is resolved using LBS based exclusively on the cellular WAN information in the enterprise IoT platform. Although this LBS location is not as accurate without GPS and Wi-Fi information, it is good enough to resolve locations inside a warehouse or distribution center and allows the Smart Label to last 6 months with good power management by the power unit.

This concludes our review of the device architectures for the Connected Cooler and Smart Label devices, including a description of the key device components and functionality in the context of the high-level solution requirements. Now, let us analyze the end-to-end architectures for these case study IoT solutions in the context of the high-level IoT architecture presented in *Chapter 2, Understanding IoT Devices and Architectures*.

Analyzing the network architectures

We will now use the basic IoT architecture presented in *Figure 2.2* in *Chapter 2, Understanding IoT Devices and Architectures*, as shown again in *Figure 8.5* here, to analyze the critical architecture decisions for the Connected Cooler and Smart Label solutions:

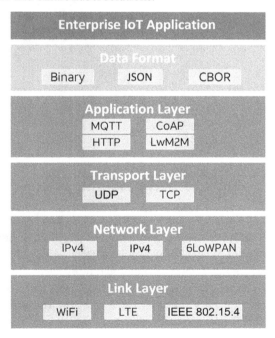

Figure 8.5 – High-level IoT protocol stack

A common theme in our review is the trade-off between low power and robust, secure communication between the IoT device and enterprise IoT application. Let us begin with the Connected Cooler solution.

Connected Cooler

Based on the business case and high-level requirements presented earlier, the Connected Cooler device is primarily powered and needs to securely report temperature and door openings for usage and product integrity data to the enterprise application, as shown in *Figure 8.6*:

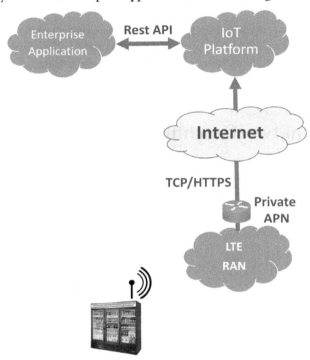

Figure 8.6 – High-level Connected Cooler network architecture

The Connected Cooler uses LTE Category 1 with 3G/2G fallback technologies with multiple frequency bands for connectivity to LTE **Radio Access Networks** (**RANs**) globally. For privacy and security, this network connection uses a private **Access Point Name** (**APN**) designated by the selected network carrier, as we reviewed in *Chapter 4, Leveraging Cellular IoT Technologies*. This private APN includes a firewall to block unwanted internet traffic from reaching the Connected Cooler device. In terms of the transport and application layer, the device sends reports using the **Transmission Control Protocol** (**TCP**) and **Hypertext Transfer Protocol Secure** (**HTTPS**), which uses **Transport Layer Security** (**TLS**) for high security in the data traffic. Using these transport and application layer protocols allows for more secure and reliable communications at the cost of higher data payloads in the reports and longer communication exchanges due to the requirement for transmission acknowledgments between the device and IoT platform. Given that the Connected Cooler is primarily powered and normally reports once a day, the more secure and reliable communications are worth the trade-off. This takes us to the data format and enterprise application layers in the Connected Cooler solution protocol stack, as shown in *Figure 8.7*:

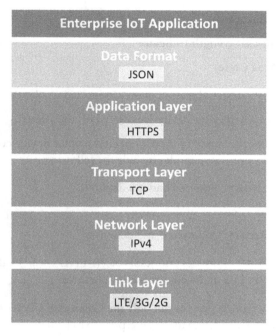

Figure 8.7 – Connected Cooler solution protocol stack

Given the concise, text-based JSON format, it was selected as the data format in the payload for the Connected Cooler HTTPS reports sent to the enterprise application. The payload values selected for the Connected Cooler JSON reports are as follows:

- **Message Type**: Indicates the type of report.
- **ICCID**: The 20-digit SIM ICCID.
- **Timestamp**: Date and time of the report.
- **Firmware Version**: The firmware version of the Connected Cooler device.
- **Temperatures**: Array of hourly average temperature reports.
- **Door**: Array of hourly door opening counts.

- **Mobile Network Data:** The cellular network information used by the Connected Cooler device. It includes the following parameters:

  - MCC (**Mobile Country Code**)
  - MNC (**Mobile Network Code**)
  - CellID (**Cell Identification**)
  - RSSI (**Received Signal Strength Indicator**)
  - LAC (**Location Area Code**)

- **GPS Location:** The GPS-derived location of the Connected Cooler device where available.

- **Power State:** Indicates whether the device is powered or unpowered.

- **Cooler State:** Indicates the mode of the Connected Cooler device concerning power and deployment at a retail store. These modes include the following:

  - **Warehouse Mode:** Battery-powered and not deployed
  - **Operating Mode AC:** Powered and deployed
  - **Operating Mode Battery:** Unpowered and deployed

- **WiFi Data:** Array of MAC addresses and RSSIs of Wi-Fi APs seen by the Connected Cooler device, which is used for LBS in the enterprise IoT platform.

The Connected Cooler JSON reports provide not only the Connected Cooler temperature and door opening data but also the state of the device to the enterprise IoT cloud platform, which is a Microsoft Azure application that parses the JSON reports and performs the location lookup using an LBS for those devices not reporting GPS latitude/longitude locations. Once formatted, this data is pushed to the enterprise application using **Representational State Transfer Application Programming Interfaces (REST APIs)**. This is a common architecture in many enterprise IoT solutions, where the IoT platform is the endpoint for the IoT data and pre-processes the data before sending it via APIs to the enterprise application. Now, let us review the architecture of the Smart Label solution, which is like the Connected Cooler but has low power and cost constraints.

Smart Label

Based on the business case and high-level requirements presented earlier, the Smart Label is a low-cost, battery-powered device that reports on the location and status of shipments concerning opening/ tampering, temperature, and shock. The Smart Label solution architecture is shown in *Figure 8.8*:

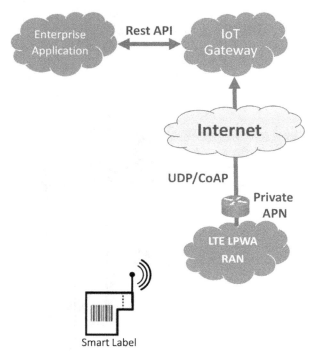

Figure 8.8 – High-level Smart Label architecture

Unlike the Connected Cooler, the Smart Label uses LTE LPWA (LTE-M and NB-IoT) technologies, as covered in *Chapter 4, Leveraging Cellular IoT Technologies,* for low-power operation with multiple frequency bands for connectivity to LTE RAN networks globally. Since neither LTE-M nor NB-IoT technology is deployed globally, the Smart Label supports both technologies. Like the Connected Cooler solution, for privacy and security, the Smart Label uses a private APN designated by the selected network carrier. In terms of the transport and application layer, as shown in *Figure 8.9,* the device sends reports using the **User Datagram Protocol (UDP)** and **Constrained Application Protocol (CoAP)**, which are lightweight connectionless protocols that do not require device acknowledgments such as TCP. With UDP, there is no overhead for maintaining the connection with the device, which saves device power by reducing the data payload and not requiring retransmissions. The drawback of using UDP and CoAP is less robust and potentially less secure communication since TLS is optional but given the low-power requirement, this trade-off is worth it. Like the Connected Cooler, the text-based JSON format was selected as the data format in the payload for the Smart Label reports to the IoT gateway and enterprise IoT application. The Smart Label data is stored and reported as objects using the **Lightweight Machine to Machine (LwM2M)** protocol, as covered in *Chapter 2, Understanding IoT Devices and Architectures.* The IoT gateway shown in *Figure 8.8* is an LwM2M server that can report to the enterprise application using REST APIs:

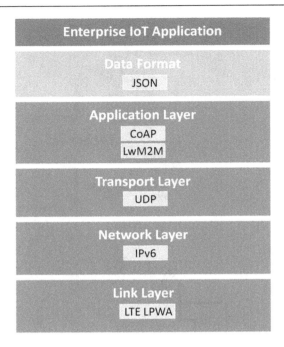

Figure 8.9 – Smart Label solution protocol stack

The Smart Label JSON report that's sent to the IoT gateway using UDP and CoAP includes the following LwM2M data objects stored on the device:

- **Device identification**: Unique IMEI of the device.

- **Report time**: Time of the Smart Label report in **Universal Time Coordinated** (**UTC**) format.

- **Trigger reason**: The reason for the report (for example, tamper, impact, temperature, movement, and so on).

- **Seal indication**: Indicates whether the shipment has been opened/tampered with.

- **Battery level**: Indicates the battery voltage level.

- **Max temperature**: Maximum temperature measured and time.

- **Min temperature**: Minimum temperature measured and time.

- **Avg temperature**: Average temperature measured.

- **Mobile Network Data**: The cellular network information used by the Smart Label device. It includes the following parameters:

  - MCC (**Mobile Country Code**)

  - MNC (**Mobile Network Code**)

- **CellID (Cell Identification)**
- **RSRP (Reference Signal Received Power)**
- **LAC (Location Area Code)**

This concludes our review of the Connected Cooler and Smart Label solutions, including the business cases, high-level requirements, and the device and network architectures. From the review of these case studies, it should be clear what trade-offs in both the device and network architectures were made to enable a low-cost and low-power solution such as the Smart Label compared to the robust, powered Connected Cooler solution. To conclude our analysis of these two case studies, we will now review the important lessons learned from these IoT solutions to provide insight into how to create a more robust enterprise IoT solution.

Lessons learned

The key lessons learned from these case studies can be categorized into three main areas:

- Piloting the solution
- Device management
- Life cycle management

One of the lessons learned from deploying both the Connected Cooler and Smart Label IoT solutions is to spend time on testing and piloting before launching the solution.

Piloting the solution

In the case of the Connected Cooler solution, we initially tested less than 50 devices in the field with Connected Cooler IoT devices retrofitted to coolers that were already deployed in the field. This initial testing was less than 6 months before the solution was launched, with over 60K shipped in the first year. With the launch, Connected Cooler devices were installed in coolers by the cooler manufacturers in their factories before shipping the coolers to retail stores. Within a few months of the product launch, we found several firmware issues on the Connected Cooler device. Most of the issues were related to incorrect sensor data and timestamps in the reports. Other issues were unexpected device reporting behavior when deployed in the retail store environments globally. Most of these issues could have been discovered in a controlled 6-to-12-month pilot of a few hundred devices that were installed in coolers by the cooler manufacturers and deployed globally. Monitoring devices in a controlled pilot would have identified many of the firmware issues and made device firmware updates easier given the small population of deployed devices.

In the case of the Smart Label solution, there were several pilots over a year with actual shipments, which helped find some key device issues impacting battery life and location accuracy. For example, the device spent a lot of time scanning LTE LPWA network technologies and frequency bands when trying to register to the cellular network, which consumed significant battery power. By implementing device logic to limit the scope of these scans, the battery life was significantly improved. In another example, the resolved location of the device using LBS was not as accurate as expected, so improvements were made in the logic for LBS in the IoT gateway to improve resolved location accuracy.

Based on these case studies, spending time testing and piloting an enterprise IoT solution is critical to finding issues before the solution is launched. Once the IoT solution has been launched, device management becomes the key to managing IoT device issues in the field.

Device management

Device management is often overlooked in many enterprise IoT solutions, but it is essential for remotely identifying and fixing IoT device issues. Once deployed, it is usually very difficult to physically access the IoT device, which makes remote device management a critical component in any enterprise IoT solution. In the case of the Connected Cooler solution, more than 50K devices were shipped globally before device firmware issues were found. Since the Connected Cooler solution included an LwM2M device management component, it was possible to identify and update devices in **Firmware Over The Air** (**FOTA**) update campaigns. This device management solution was critical in managing the Connected Cooler devices over the life cycle of the IoT solution, with over 500K devices ultimately shipped in over 60 countries. In the case of the Smart Label, where LwM2M was used for device reporting, device management was intrinsically part of the solution, which made device management easier. Given the Smart Label was very constrained in terms of power and processing capabilities, device management was primarily used for device identification, configuration changes, and health status and not for firmware updates. Using the LwM2M protocol, the data objects on the Smart Label could be easily configured when the device reported to the IoT gateway.

The other aspects of device management are device watchdogs and self-recovery methods implemented on the IoT device. Since it is impossible to find every error condition an IoT device could encounter in the field, even with an extensive pilot phase, the device needs to have a self-recovery mechanism when it is no longer functioning for whatever reason. This could be as simple as a timer that resets the device periodically, or it could be more complex with either a software or hardware watchdog component implemented on the device to detect an error condition and then perform a reset/reboot or power cycle of the device for recovery. In the case of the Connected Cooler solution, several recovery timers and software watchdogs were implemented on the device to enable self-recovery in the field. These recovery mechanisms improved the overall stability of the Connected Cooler devices in the field without active remote device management.

Overall, device management should be considered a critical part of any enterprise IoT solution, which is also part of the solution life cycle management.

Life cycle management

In addition to device management, there are other aspects to cellular IoT solution life cycle management, as we will cover in *Chapter 9, Managing the Cellular IoT Solution Life Cycle*, including SIM and cellular data management, as well as device certifications. With the manufacturing, provisioning, deployment, monitoring, and sunset of IoT devices as part of a cellular IoT solution, it is critical to manage cellular data usage on devices, especially concerning roaming, where data charges could be quite high. Many cellular carriers offer SIM data management applications, which allow enterprise customers to activate or deactivate device SIMs and monitor data usage, and even set rules for data overages to avoid unnecessary charges. This SIM management is especially important in IoT solutions with thousands of devices deployed globally, where they could be roaming on foreign carriers, which could have higher cellular data charges. In the case of the Connected Cooler solution, the IoT devices were installed and tested by the cooler manufacturers before being shipped globally. During this testing and installation, it was important to avoid any data charges, so the network carrier set a reasonable data threshold for this pre-deployment phase in the Connected Cooler solution before the device SIMs were activated for data billing on the SIM management platform. Once deployed, many of the Connected Cooler devices exceeded the expected data usage due to a high number of reporting events, so in some cases, the SIMs were deactivated to avoid excessive data charges. Overall, the SIM management of the Connected Cooler device SIMs was critical in managing the life cycle's recurring data costs of the IoT solution.

There are typically two types of certifications for IoT devices in an enterprise IoT solution. These are the network carrier and country regulatory certifications. Most network carriers require certifications for IoT devices operating on their domestic network to validate their performance using the carrier SIM. These certifications typically require third-party lab testing of the RF performance of the device in the network carrier frequency bands, along with network carrier testing of the device behavior on their live network. There is time and cost associated with this testing, which needs to be considered when planning the IoT solution. Country regulatory certifications are the second type of device certification required for IoT devices deployed in a global IoT solution. Most countries require regulatory approvals for electronic devices that are deployed in their country. With IoT solutions that are deployed in several countries, the time and cost of these approvals can be daunting. In the case of the Connected Cooler global solution, both variants of the Connected Cooler device were certified with a network carrier, which enabled global roaming on the network carrier roaming partners. For country regulatory certifications, the specific countries were identified at the start of the project and budgeted in the enterprise solution; however, other countries were added to the project later, which required additional unplanned budget and time. Overall, for global IoT solutions, it is important to identify the countries where devices will be deployed and plan/budget accordingly to avoid any surprises later.

In summary, the main lessons learned from the Connected Cooler and Smart Label IoT solution case studies are as follows:

- Spend 6 to 12 months piloting and testing your enterprise IoT solution in the actual field deployment environments before launching
- Ensure that device management is included in your IoT solution to support device identification, monitoring, and FOTA support

- As a part of device management, ensure that hardware/software watchdogs are implemented on the IoT device

- As a part of life cycle management, plan on certifying the IoT device with the selected network carrier and completing the regulatory certifications in the countries where the enterprise IoT solution will be deployed

- Ensure that SIM and cellular data management is part of your IoT solution life cycle management plan to help manage recurring data charges

This concludes our review of the Connected Cooler and Smart Label IoT case studies. Now, let us summarize the key points in our discussion.

Summary

In this chapter, we presented two real-world enterprise IoT solution case studies: the Connected Cooler and Smart Label solutions. We presented the business cases, high-level requirements, device architectures, and network architectures for each solution and highlighted the architectural decisions driven by the cost, power, security, and global coverage requirements for each solution. These architectural trade-offs should provide a good framework for the development of any enterprise IoT solution. We concluded by reviewing the lessons learned from these IoT solutions, which centered around piloting/testing, device management, and solution life cycle management. Our review of these two case studies should provide insight and guardrails for a more robust enterprise IoT solution.

In the remaining two chapters, we will cover the cellular IoT solution life cycle and explore some of the new IoT technologies, trends, and business models that could be incorporated into your enterprise IoT solution.

Part 3: Cellular IoT Solution Life Cycle and Future Trends

In this final part of the book, we will review the full cellular IoT solution life cycle for an enterprise IoT solution and discuss some emerging technology trends in IoT that will impact new IoT solutions over the next 10 years.

This part contains the following chapters:

- *Chapter 9, Managing the Cellular IoT Solution Life Cycle*
- *Chapter 10, Looking at the Road Ahead*

Managing the Cellular IoT Solution Life Cycle

So far in this book, we have reviewed the value of the IoT, enterprise IoT markets, IoT device and network architectures, IoT wireless technologies, IoT security, and IoT solution case studies with a focus on the cellular IoT. Given that enterprise IoT solutions are expected to perform in the field for many years, it is important to plan the full cellular IoT solution life cycle as part of your enterprise IoT strategy. It should be clear that a full cellular IoT solution is complex with several components, including the IoT device, wireless connectivity, cellular network, IoT platform, and enterprise application – the chosen system architecture components impact the performance of the overall cellular IoT solution. Most businesses do not have the expertise to develop each of these disparate components, and unlike a consumer product life cycle, with an enterprise IoT solution, developers need to interact with, monitor, and maintain the solution over its full life cycle from designing and planning to sunset. In this chapter, we will discuss these challenges in the context of an overview of the cellular IoT solution life cycle. We will conclude with some best practices for managing the life cycle of your own enterprise cellular IoT solution. We will cover the following topics in this chapter:

- The challenges of IoT life cycle management
- The cellular IoT life cycle
- Best practices

Let us explore some of the main challenges in developing and managing your enterprise IoT solution further and how to address them, before reviewing the cellular IoT life cycle.

The challenges of IoT life cycle management

As just introduced, the primary challenges in developing and managing an enterprise IoT solution are implementation complexity and full life cycle support. In terms of complexity, an enterprise IoT solution has multiple technology components from the IoT device to the enterprise application, which are typically sourced from various hardware, software, and service vendors. Moreover, many businesses rely on their IT departments to develop and manage their enterprise IoT solution, which are not typically resourced to handle the complexity of the development and interworking of the IoT solution technology components and manage the full IoT solution life cycle. This is typically addressed by working with a third-party system integrator service company that is well-versed in the IoT technology ecosystem and can develop, deploy, and monitor a full enterprise IoT solution based on the business requirements. This outsourcing can be expensive over the entire IoT solution life cycle, especially with new requirements that may be added in the future, and can lead to vendor *lock-in* when the IoT solution becomes dependent on specific suppliers, which reduces the flexibility of the overall IoT solution. An alternative approach in which a business can maintain better ownership of the enterprise IoT solution life cycle is to outsource the more complex components such as the IoT device, wireless service, and IoT cloud platforms while focusing on the unique, value-added enterprise IoT application. Let us look at the specific IoT solution technology components that are commonly outsourced in more detail, starting with the IoT device.

IoT devices

For most enterprise IoT solutions, it is best to use an off-the-shelf IoT device, potentially with customizable firmware to meet the enterprise solution requirements. With all the cellular IoT devices available on the market that we reviewed in *Chapter 6, Reviewing Cellular IoT Devices with Use Cases*, it rarely makes sense to build a custom IoT device that requires hardware or software design resources, significant **Non-Recurring Engineering** (**NRE**) expense, and, typically, about a 1-year design cycle. In most enterprise IoT solutions, the value differentiation is normally the unique enterprise application or data and not a custom IoT device. Many of the IoT device **Original Equipment Manufacturers** (**OEMs**) allow for configuration and firmware customization to support a broad range of enterprise IoT solutions, which removes the impetus to develop a custom IoT device. If there is not an off-the-shelf IoT device that meets the IoT solution business requirements, there are several **Original Device Manufacturers** (**ODMs**) that specialize in building custom, white-labeled devices. In the course of the past year, certain cellular module suppliers, such as Quectel and Telit, have begun to offer ODM services for IoT device development so that IoT solutions can use their models. This brings us to the selection of cellular wireless network carriers.

IoT wireless services

As we covered in *Chapter 3, Introducing IoT Wireless Technologies*, several licensed and unlicensed wireless technologies could be used in an enterprise IoT solution, with licensed cellular service providers such as AT&T, Verizon, and T-Mobile offering fully managed public and private LTE and 5G wireless IoT services. Using unlicensed wireless technologies typically requires the business to support the wireless network over the life cycle of the enterprise IoT solution, which can incur significant operational support expenses. To help manage the ongoing operational expense of an enterprise IoT solution, it is best to use a cellular network carrier for the wireless service, as the maintenance, support, and upgrades of these wireless networks over the IoT solution life cycle is its core business and included in the cost of the IoT data plans. Many of the network carriers also provide other services such as device and data management, which help with monitoring and maintaining devices over the IoT solution life cycle. As discussed in *Chapter 7, Securing the Internet of Things*, using cellular network technology with a full-service network carrier also helps ensure the life cycle security and privacy of data in your enterprise IoT solution, which is not inherent to an unlicensed wireless service. Let us now look at the IoT platform, which is another enterprise IoT solution technology component that is commonly outsourced.

IoT platforms

As described in *Chapter 2, Understanding IoT Devices and Architectures*, the IoT platform is the ingestion point for the enterprise IoT solution data before it is sent to the enterprise application. The IoT platform provides the generic functionality of the enterprise IoT solution to allow for differentiation in the actual enterprise application. An IoT platform ingests the enterprise IoT data and may also perform data analytics, integrate edge computing, and store the data before forwarding it to the enterprise application using API calls. The most common IoT cloud platforms used today are **Amazon Web Services** (**AWS**) and Microsoft Azure. Both IoT platforms allow the rapid development of applications specific to your enterprise IoT solution and typically do the *heavy lifting* for the IoT solution in terms of data analytics, cloud computing, and data manipulation. Both the AWS and Azure IoT platforms can support nearly all enterprise IoT solution requirements and have become standard components of many IoT solutions. When selecting the IoT platform for your enterprise IoT solution, it is important to consider the following questions:

- Does the platform allow for multiple custom portals and levels of user hierarchies?
- Is the platform compatible with existing enterprise applications that have already been developed?
- How much effort is required to integrate the IoT platform with existing solutions?
- Does the platform allow for both on-premises and public cloud-based hosted environments?
- Does the platform use industry-standard protocols?
- How is security managed in the platform?
- Does the cost model for the platform allow for a modular *pay-for-what-you-use* service?

In summary, there are more and more system integrators in the IoT ecosystem providing architecture components across the full IoT solution stack, but the implementation complexity of an enterprise IoT solution can be addressed best by only partnering with third-party providers on the more complex system architecture components such as the IoT device, wireless connectivity, and IoT platform while keeping ownership over the differentiated value of your enterprise IoT solution.

The second main challenge in managing an enterprise IoT solution is the requirement for full life cycle support. Nearly all enterprise IoT solutions are expected to perform at scale over many years, which means businesses need to monitor, maintain, upgrade, and optimize the IoT solution over several years. This long-term support needs to be anticipated in the initial planning and design phase of the IoT solution, which includes planning a robust device management solution, as covered later in this chapter. This full life cycle support planning needs to anticipate each stage of the enterprise IoT solution life cycle shown in *Figure 9.1* here:

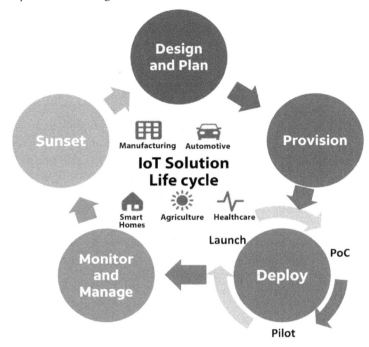

Figure 9.1 – IoT solution life cycle

Let us review each stage of the IoT solution life cycle in more detail to provide a framework to guide your organization through this challenging process.

The cellular IoT life cycle

With the scale, complexity, and longevity of a typical enterprise IoT solution, it is critical to consider each stage of the IoT solution life cycle at the very beginning of the designing and planning stage.

Designing and planning

This is the first and most important stage in the life cycle, as it sets the foundation for the success of the overall enterprise IoT solution. In this stage, the IoT business case and its various stakeholders, business lines, and customers are defined. During this business case review, developers should not only consider the IoT solution requirements but also the requirements for each of the following four stages leading to solution sunset. For enterprise IoT solutions that build upon existing enterprise systems, it is also important during this stage to understand how to best bridge the incumbent products and processes with the new enterprise IoT solution requirements. This leads to an evaluation of which architecture components to reuse and whether to build or buy the remaining system components depending on internal expertise and cost. Part of this evaluation is surveying the available cellular IoT devices, LTE and 5G wireless carrier services, and IoT platforms that can work seamlessly with your enterprise application. While the business cases and associated IoT solution architectures vary by use case, some of the high-level questions to answer in this evaluation are as follows:

- What data is needed in the IoT application?

- How is this data collected, stored, and reported?

- How much data is generated?

- How is data secured end-to-end?

- What data analytics are needed to support business insights?

- What enterprise systems need to be connected to the IoT solution?

The answer to these questions impacts all the architecture component decisions in the enterprise IoT solution. This brings us to the second stage of provisioning, which includes the IoT device, wireless network, and IoT cloud platform or application.

Provisioning

Provisioning is a key stage in the IoT solution life cycle in which the IoT device identity, cloud endpoint, platform security credentials, and wireless network credentials are configured before deployment. This encompasses device, network, and IoT platform provisioning and includes onboarding or associating the device identity to the IoT platform and enterprise application, as well as configuring the device with the associated IoT platform security credentials and data reporting endpoint. When preparing the device for connection with the enterprise application, the device is also typically paired with the asset (for example, the factory equipment, cooler, shipping container, or fleet vehicle) that it is monitoring. From a cellular network perspective, device provisioning includes the activation of the SIM on the correct carrier account and configuring the device with the correct APN for network access. Most network carriers also require the device IMEI information to keep a record of the IoT devices deployed on their network. This provisioning process is typically done in the manufacturing process before the shipment or deployment of the device as part of the enterprise IoT solution. Some of the IoT platform and cellular network providers offer a **Device Provisioning Service (DPS)** for quickly getting IoT devices connected and configured to their platform, including options for zero-touch provisioning, whereby the IoT device automatically connects to the IoT platform, which then applies the proper configuration to the device based on the IoT solution requirements. Using a DPS streamlines the number of manual steps required in the device provisioning process, saving time during manufacturing, and should be discussed as an option with your IoT platform and cellular network providers. This brings us to the deployment phase of the IoT solution life cycle.

Deployment

The deployment phase of the IoT solution life cycle is the most time-consuming stage, as it normally includes a **Proof of Concept** evaluation and pilot over several months before the full launch. As mentioned in the lessons learned from *Chapter 8, Implementing an IoT Solution with Case Studies*, it is important to evaluate your enterprise IoT solution not only in a basic Proof of Concept but also in a pilot over a few months to identify any issues and assess the feasibility of the IoT solution. This would potentially necessitate another Proof of Concept and pilot before launch. The point is that identifying issues in the IoT solution before a full launch saves both time and money in the long term by not having to replace or remotely update devices in the field that are normally inaccessible. The deployment phase of an enterprise IoT solution is more complex than traditional products, as there are typically many stakeholders involved, such as the device OEMs, system integrators, and internal or external customers that may need access to the solution over its life cycle, which is a factor that needs to be managed as well. Once launched, the management and scalability of the IoT solution become the key to overall success, which brings us to the fourth stage in the IoT solution life cycle – *monitoring and management*.

Monitoring and management

During the monitoring and management phase of the IoT solution life cycle, solution stakeholders will remotely monitor the health of the deployed IoT devices and should be able to remotely update the configuration and firmware on the devices as needed. This is the critical device management function we described in the lessons learned from *Chapter 8, Implementing an IoT Solution with Case Studies*. Since most IoT devices deployed in an enterprise IoT solution are inaccessible or difficult to access, having the ability to remotely troubleshoot and update devices is critical to the success of your enterprise IoT solution. In addition to device monitoring and updating, managing the device's cellular data usage is critical for cost management in the IoT solution life cycle. Although you can estimate the expected data usage on individual IoT devices based on the reporting payload and intervals, there will always be cases where devices consume more data than expected. Given this, many cellular network carriers offer services or applications that enable their IoT customers to manage their IoT data usage by changing data plans or establishing rules for SIM activation and deactivation.

Several IoT device management platforms are offered by IoT solution ecosystem providers, such as cellular network carriers, IoT system integrators, IoT device OEMs, and cellular module suppliers. The basic requirements for any device management platform to adequately support an enterprise IoT solution are as follows:

- Support for remote IoT device configuration and firmware or software updates (**Firmware Over The Air** (**FOTA**) or **Software Over The Air** (**SOTA**))
- Support for device health monitoring
- Support for RESTful APIs to pull data into the enterprise IoT platform or application

- Support for the bulk management of devices by group

- Support for auto-scaling to handle the growth of an enterprise IoT solution

- Interoperability support for the IoT devices in the IoT solution using industry-standard protocols such as LwM2M, MQTT, and HTTP

- Support for best-in-class security

Monitoring and managing the enterprise IoT solution is a critical ongoing operational task over the entire solution life cycle. It not only enables quicker and more cost-effective delivery of support at scale but also creates a better end user or partner deployment experience. The last stage of the IoT solution life cycle, which is often overlooked, is the sunset phase where IoT devices are removed from service and new devices are onboarded.

Sunset

In the final stage of your enterprise IoT solution, it is important to easily be able to decommission legacy IoT devices and onboard new ones, making it simpler for internal and external customers to transition from one solution to another. This provides the impetus for your customers to continue investing and using the enterprise IoT solution and builds loyalty. The sunset process typically involves the removal of the device from the device management platform and IoT platform or application, as well as the deactivation of the device SIMs with the cellular network carrier. When possible, the IoT devices are recovered and repurposed for other solutions. As an example, when cellular network carriers sunset their 2G or 3G networks, the 2G or 3G IoT devices that were part of many enterprise IoT solutions are no longer able to connect to a network. The owners of these IoT solutions are required to follow a sunset process in which devices are decommissioned and replaced with newer LTE devices. A critical part of this process is coordinating with the customers impacted and informing them about the sunset process when transitioning to newer IoT devices. Overall, it is important to include the sunset stage in your enterprise IoT life cycle planning to make transitions to new IoT products more efficient for both internal and external customers.

We have reviewed the five stages of an IoT solution life cycle from designing and planning to sunset, and it should now be clear why each stage is important in the success of your enterprise IoT solution. Most enterprise IoT projects never go beyond the Proof of Concept stage in deployment primarily due to poor business case reviews, lack of stakeholder buy-in, and insufficient requirements gathering in the planning stage. To help ensure the success of your enterprise IoT solution, let us now look at some best practices for each stage of the IoT solution life cycle.

Best practices

In this section, we will cover some of the best practices for each stage of the IoT solution life cycle, starting with the designing and planning stage.

The designing and planning stage

As described earlier, this is the most critical phase in the enterprise IoT solution life cycle, as it sets the foundation for success in the follow-up phases of provisioning, deployment, and management. There are a few best practices to consider at this stage to help ensure the success of your IoT solution. Here are three common factors that contribute to the failure of many enterprise IoT solutions:

- IoT solutions take much longer than expected to launch
- The underlying enterprise's cultural and organizational change is underestimated
- The technical skills required by the IoT solution are often not available in the organization

In terms of time-to-market, it is important to realize that most enterprise IoT projects take much longer than planned to launch, with the average time-to-market being 18-24 months. There are many reasons for this delay, including technical issues and a lack of buy-in from stakeholders. In this context, a best practice is to reevaluate the expected revenue and cost savings from the solution and plan for project delays. Another best practice is to select IoT partners that have experience with similar IoT solutions, which can help to avoid business and technical pitfalls on the journey to launching your enterprise IoT solution.

Given the disruptive nature of a new enterprise IoT solution that transforms business models in the context of digital transformation, it is important to understand the significant cultural and organizational changes required in the business to support this transformation. This requires both the alignment of the workforce with the new IoT business model and support across many different departments within the business. This typically leads to the break-up of work silos and can result in significant opposition to change. Moreover, the complexity of successfully implementing an IoT solution requires new project management methodologies such as agile development, which conflicts with more traditional, less flexible project management styles such as the waterfall technique. Agile development is an iterative approach to project management and solution development in which teams deliver work in small increments to respond to change quickly, as opposed to the *big bang* launch associated with traditional waterfall development methodologies. A best practice to address this disruption to legacy development techniques is to get early buy-in for the IoT solution from senior management and stakeholders and introduce the agile development processes within the organization.

The development of an end-to-end enterprise IoT solution requires a broad range of skills, including hardware and software design, cloud architecture design, and platform integration experience, which most organizations lack. As a best practice, it is important to work with the IoT partner ecosystem to address this gap in skills until the new IoT skills can be developed internally within the organization. As a part of this IoT skill development, it is important to identify the IoT skill gaps and cross-train the workforce accordingly, potentially working with IoT experts from different domains in the IoT ecosystem. Addressing this IoT skills gap enables businesses to better develop and manage their enterprise IoT solution without as much reliance on partners.

Given the importance of the design and planning stage in setting the foundation for your enterprise IoT solution, following these best practices will help ensure a successful launch. This brings us to a review of some best practices in the provisioning stage.

Provisioning

In the provisioning stage, the IoT device is onboarded to the IoT platform or application, and the cellular network SIM is activated to allow the device to connect to the LTE or 5G network. There are multiple steps in the device provisioning process during manufacturing, so it is important to streamline the process as much as possible to save time and avoid mistakes. The key areas where this process can be streamlined are as follows:

- Cellular network SIM provisioning
- IoT platform or application onboarding

Let us now look at these two aspects of provisioning and some best practices for each. In terms of cellular network provisioning, it is important to coordinate with your cellular network carrier during the setup of the IoT account to support the features and security required of the IoT solution, which include the following:

- Voice or data support
- The data plan setup
- SMS support
- International roaming support
- The VPN configuration for network security

As described in *Chapter 4, Leveraging Cellular IoT Technologies*, these cellular network features are provisioned on the SIMs associated with the enterprise IoT account. When the IoT device is provisioned during manufacturing, the SIM is activated to allow network connectivity. A best practice in the SIM activation process is to establish a data usage threshold when the cellular IoT data billing starts, which allows the device to be tested during manufacturing without incurring data charges. This data threshold is typically around 20 KB and allows the device to connect to the IoT platform for onboarding and end-to-end testing.

The IoT platform or application onboarding is the provisioning step in which the IoT device is configured with the IoT platform security credentials and reporting endpoints for end-to-end testing. This is typically a manual process where the device identities and security credentials are uploaded to the platform or application for onboarding. As described earlier, the IoT platform and network carriers may offer a DPS to get IoT devices automatically connected and configured to the IoT platform. With this service, the IoT device automatically connects or bootstraps to the IoT platform during factory provisioning, which then applies the proper configuration to the device. This is a best practice for streamlining the IoT platform provisioning process, saving time and minimizing device configuration errors in manufacturing.

Let us now review some best practices in the deployment stage of the IoT life cycle.

Deployment

As described earlier, the deployment phase is the most critical phase of the IoT solution life cycle, as it takes significant time and resources to complete potentially multiple iterations of IoT solution Proofs of Concepts and pilots. This is the main reason most enterprise IoT solutions do not proceed past the Proof of Concept or pilot phase, so it is important to set expectations with senior management and get stakeholder buy-in on the IoT solution in the design and planning stage. The purpose of a Proof of Concept is to assess the feasibility of your enterprise IoT solution in scenarios that match the actual IoT use case. The main purpose of a Proof of Concept is to identify interoperability and technical issues in the IoT solution that could manifest at any layer of the IoT solution architecture, including the device, network, or application layers. These issues typically occur in the interface between different systems in the IoT solution architecture, such as the connection between the legacy enterprise application and the IoT cloud platform. As another example, many enterprise IoT solutions require IoT devices to interface with legacy industrial equipment using older machine protocols, which requires the device to implement a protocol translation to capture and report the data to the IoT platform or application. This typically requires a specialist with expertise in protocol translation to apply this to the device, which takes time during the Proof of Concept evaluation. As a best practice to reduce interoperability issues, it is important to design your enterprise IoT solution on a standardized ecosystem for your IoT use case. Using the standard IoT device and architecture protocols discussed in *Chapter 2, Understanding IoT Devices and Architectures*, will help reduce interoperability issues in your IoT solution. With a successful Proof of Concept, the next step is to evaluate the Proof of Concept at scale in a pilot, where you can further assess the features of your IoT solution in real-world scenarios. A pilot with a few hundred IoT devices builds on the Proof of Concept in terms of finding interoperability and technical issues that may have been missed in the small-scale Proof of Concept. As discussed in the lessons learned in *Chapter 8, Implementing an IoT Solution with Case Studies*, it is a best practice to run the pilot over at least 6 months to adequately vet the IoT solution at scale.

As reviewed in *Chapter 7, Securing the Internet of Things*, it is important to follow best practice end-to-end security guidelines throughout the overall IoT life cycle, and especially all phases of the deployment stage. With time-to-market pressures on the launch of your enterprise IoT solution, it is tempting to overlook the security of a **Minimum Viable Product** (**MVP**), but this leads to numerous security threats from attackers for a commercial launch at scale. Many of these security threats can be minimized using the device security guidelines in *Chapter 7, Securing the Internet of Things*, and using secure network-based VPNs over the public cellular network. In summary, a best practice is to implement these security guidelines in all phases of deployment from the Proof of Concept to commercial launch.

With a successful Proof of Concept and pilot leading to the commercial launch of your enterprise IoT solution, let us now review the best practices in monitoring and managing the IoT solution in the field.

Monitoring and management

With the commercial launch of your enterprise IoT solution, it is now important to remotely manage the health of IoT devices and remotely update the configuration and firmware on these devices as needed. This is where the importance of having a device management solution described earlier in this chapter is realized. Even with a successful Proof of Concept and pilot, device issues that require remote updates will emerge for large-scale deployments. As such, the device management platform should be able to provide analytics on device health and support firmware or software updates at scale. Since these device updates use cellular data, a best practice is to coordinate with the device OEM on delta image updates, as opposed to updating the full firmware or software image, to reduce data usage. As discussed in the device management lessons learned in *Chapter 8, Implementing an IoT Solution with Case Studies*, it is a best practice, when possible, to use IoT devices with hardware and software watchdog features to enable self-monitoring and recovery in cases where the device is not able to connect to the device management platform.

We have reviewed some of the best practices in the IoT solution life cycle to help ensure the success of your enterprise IoT solution. These are summarized as follows:

- Plan on an 18-24 month time-to-market launch and get senior management and key stakeholder buy-in accordingly
- Plan to align the workforce with the implementation of a new and disruptive enterprise IoT solution and get senior management buy-in
- Plan for a skills gap when developing the new IoT solution and coordinate with the IoT partner ecosystem to fill this gap until the internal workforce has been developed and trained as required
- Coordinate with your selected cellular network carrier on the setup of your IoT account to support the features of your IoT solution
- Coordinate with your selected cellular network carrier on SIM provisioning and IoT data plans
- Coordinate with your IoT platform provider or cellular network carrier on their device provisioning service to streamline device provisioning
- Design your enterprise IoT solution on a standardized ecosystem for your IoT use case and use standard IoT device and architecture protocols to minimize interoperability issues
- Plan on a pilot of your enterprise IoT solution over at least 6 months to adequately vet the IoT solution at scale
- Plan on end-to-end security during all stages of the IoT solution life cycle
- Include a device management solution in your IoT solution that can monitor device health and remotely update devices in the field

Let us now summarize our review of the cellular IoT solution life cycle, including the challenges and best practices to enable a more robust enterprise IoT solution.

Summary

In this chapter, we presented the challenges within enterprise IoT solution life cycle management, which can be categorized as implementation complexity and full life cycle support. We then presented an overview of the five stages of the IoT solution life cycle, including design and planning, provisioning, deployment, monitoring and management, and sunset, with a description of the key activities and considerations at each stage. Finally, we reviewed the best practices for the design and planning, provisioning, deployment, and monitoring and management stages to help ensure the success of your enterprise IoT solution.

In *Chapter 10*, *Looking at the Road Ahead*, we will cover new and emerging technology trends in and business models for enterprise cellular IoT solutions.

10
Looking at the Road Ahead

In the first six chapters of this book, we reviewed the importance of the IoT in many key markets, followed by an analysis of the IoT devices, networks, wireless technologies and architectures, with a focus on cellular IoT solutions. In *Chapter 7, Securing the Internet of Things*, we described the importance of security in your IoT solution, along with some practical security guidelines to follow. In *Chapter 8, Implementing an IoT Solution with Case Studies*, we analyzed two real-world cellular IoT solution case studies with some key lessons learned that can be readily applied to your enterprise IoT solution. We concluded in *Chapter 9, Managing the Cellular IoT Solution Life Cycle*, with an examination of the cellular IoT solution life cycle and some practical best practices to help ensure the success of your enterprise IoT solution. In this chapter, we will look 10 years ahead and discuss emerging cellular IoT technologies and top IoT trends and markets.

As we described in *Chapter 1, Transforming to an IoT Business*, it is expected that more than 15 billion enterprise IoT devices will be connected by 2029 (*Source: Gartner*), increasing operational efficiency, saving costs, and driving new business models in many key markets. Even though the IoT has been around for more than 20 years, we are still at the very early stages of IoT market growth, especially with new cellular technologies such as LTE LPWA networks enabling new and innovative use cases. As discussed in *Chapter 1, Transforming to an IoT Business*, the enterprise market is where the IoT is having the biggest impact, with double-digit compound annual growth in most market segments, which is the impetus for this book. In this final chapter, we will examine some emerging cellular technologies that help enable new enterprise IoT solutions, along with the top IoT technology trends and markets. We will conclude with some guidance for getting started on your enterprise IoT journey with some best practices to follow on that journey. We will cover the following topics in this chapter:

- Emerging cellular IoT technologies
- IoT trends and business models
- Getting started on your IoT journey
- Best practices in developing and launching your IoT solution

Let us first explore some of the new and emerging cellular technologies that can benefit your enterprise cellular IoT solution.

Emerging cellular IoT technologies

As discussed in *Chapter 5, Validating 5G with the IoT*, 5G is the next generation cellular technology to follow LTE with the initial 5G carrier deployments being **Non-Standalone (NSA)** and using the existing LTE core network. As we reviewed in *Chapter 5, Validating 5G with the Io*, to realize the full benefits of 5G in IoT solutions with low latency, high reliability, and high peak data rates tailored to the IoT application, the 5G carrier deployments need to be migrated to the **Standalone (SA)** architecture. In parallel with the current 5G network carrier deployments in the licensed 5G spectrum, there are a growing number of both LTE and 5G private network deployments in the **Citizen Band Radio Service (CBRS)** bands from 3.5 GHz to 3.7 GHz, also known as LTE band 48. Let us now look at private CBRS network deployments in the context of enterprise IoT solutions.

Private LTE and 5G CBRS networks

Prior to 2019, the CBRS spectrum was only accessible to the US federal government and satellite services, but this spectrum is now partially available for commercial use and has been used in an increasing number of private LTE and 5G IoT solution deployments. By making this spectrum available for commercial use, the FCC implemented a tiered shared spectrum model as shown in *Figure 10.1* here:

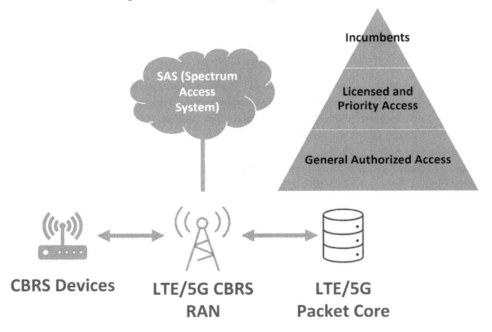

Figure 10.1 – CBRS network architecture with a shared spectrum

In this tiered shared spectrum architecture, access to the CBRS spectrum is managed by a **Spectrum Access System (SAS)** administrator designated by the FCC. There are three tiers of CBRS band access, which are the incumbent users at the top, licensed and **Priority Access Licenses (PALs)** in the middle, and **General Authorized Access (GAA)** licenses at the bottom. When an enterprise wants to use the CBRS spectrum, it must contact a SAS administrator and indicate where they want to deploy the CBRS access points for its private network. The SAS administrator then indicates whether the spectrum is available and specifies the channels and appropriate transmit power for the CBRS devices.

A private LTE or 5G network using the CBRS is like a Wi-Fi network in that it is a dedicated enterprise network with access points that can cover large areas such as that of a warehouse or factory. There are several reasons an enterprise would consider a private LTE or 5G IoT network over Wi-Fi or public cellular network, which include the following:

- Better coverage and data throughputs for enterprise locations, especially when compared to Wi-Fi
- Better privacy and security since the IoT data never leaves the enterprise network
- Better flexibility in customizing and scaling the network to meet unique enterprise IoT requirements over the full IoT solution life cycle
- Better **Quality of Service (QoS)** and data prioritization compared to Wi-Fi

With these benefits, a private network is well suited to supporting the latency, data throughput, and edge computing requirements of most enterprise cellular IoT solutions. There are an increasing number of private networks, especially in the following areas where enterprise IoT solutions are deployed:

- Hospitals
- Retail stores
- Warehouses
- Factories
- Construction sites
- Farming operations

A private LTE or 5G network allows an enterprise to own, customize, and scale its IoT network over the life cycle of the enterprise IoT solution, but there are some important considerations for selecting a private cellular network over a public network. One of the most important considerations is the cost not only of the purchase and installation of the private LTE/5G infrastructure but also the monthly charge for CBRS spectrum usage. An enterprise could either directly lease a dedicated CBRS spectrum on its own or lease the spectrum from tier 2 CBRS spectrum holders that own a PAL, such as AT&T, Verizon, and Nokia. Although using a dedicated spectrum can deliver better performance, using a shared spectrum from tiers 2 and 3 can provide a more affordable solution. The total cost of ownership for a private LTE or 5G solution, including the devices, network infrastructure, installations, and spectrum, needs to be factored into the enterprise IoT solution's expected **Return on Investment**

(**ROI**). Another important consideration is a CBRS device ecosystem where fewer IoT devices support the CBRS frequency bands to meet your enterprise IoT solution requirements. Moreover, if the IoT device also needs to roam on public LTE/5G networks, it also needs to support the technology and frequency bands of the selected public carrier and be allowed to roam on that network. Even with these considerations, private networks are increasingly adopted to meet a wide range of enterprise IoT connectivity needs.

As we reviewed in *Chapter 4, Leveraging Cellular IoT Technologies*, **Integrated Universal Integrated Circuit Cards (iUICCs)** and **Embedded Universal Integrated Circuit Cards (eUICCs)**, also termed iSIMs and eSIMs respectively, are two emerging cellular technologies that enable an expanded ecosystem of cellular IoT connectivity options. Let us review how these technologies for accessing a cellular LTE or 5G network can benefit enterprise IoT solutions.

iUICC and eUICC SIM technologies

As described in *Chapter 4, Leveraging Cellular IoT Technologies*, the iUICC or iSIM, introduced in 2020, is where the SIM storing the cellular carrier profile is embedded within a **Tamper-Resistant Element** (**TRE**) on the device's cellular chipset, eliminating the need for any discrete SIM in the device. This is now the SIM used in many smartphones. In the context of the IoT, this reduces the size and complexity of IoT devices while also saving power, which is important for low-cost-constrained IoT sensor devices. The iSIM can also be an eUICC or eSIM, which supports multiple carrier profiles and allows remote carrier provisioning of the SIM profile. This means the network carrier profile can be remotely switched in the device without physically replacing the SIM in the IoT device. An iSIM offers a few advantages in the context of an enterprise IoT solution, including the following:

- Since the iSIM is *invisible* inside the device, it is easier to configure and connect IoT devices without the need to physically handle and manage physical SIMs

- The iSIM is more secure than traditional SIMs in that they cannot be stolen or removed

- With the eSIM feature, it is easy to switch cellular network carriers and activate trials to test a new carrier, which also reduces transition costs

- With an eSIM, it is possible to remotely change carriers and update wireless data plans at the end of a carrier contract

Most **Mobile Network Operators** (**MNOs**) such as AT&T, Verizon, T-Mobile, and Vodafone, as well as **Mobile Virtual Network Operators** (**MVNOs**) such as KORE and Aeris Communications, offer eSIM options to enable your enterprise IoT solution. Going further with advancements in SIM technology following the **Global System for Mobile Communications Association** (**GSMA**) IoT security guidelines, the newest SIMs can also function as the IoT device root of trust with a specification called **IoT SIM Applet for Secure End-2-End Communication** (**IoT SAFE**). Let us now look at how IoT SAFE can help secure your enterprise IoT solution.

IoT SAFE

Using the SIM for cryptographic credential storage, IoT SAFE provides a methodology to secure IoT data using the trusted SIM as opposed to potentially less secure or less trustworthy elements on the IoT device. The IoT SAFE applet is compatible with all SIM form factors, including the eSIM and iSIM described earlier. For constrained IoT sensor devices with low compute capabilities, IoT SAFE can offload crypto processing from device processing components such as the **Microcontroller Unit** (**MCU**), which can reduce the complexity and cost of a device. From a logistics perspective, IoT SAFE enables *zero-touch* cloud service provisioning of millions of IoT devices, as a device with an IoT SAFE SIM can automatically be bootstrapped and authenticated, typically using **Transport Layer Security** (**TLS**) or **Datagram Transport Layer Security** (**DTLS**) with the enterprise IoT platform for onboarding and configuration. Given the SIM is remotely manageable, the IoT SAFE applet can also be remotely provisioned and maintained. With IoT SAFE, many of the IoT device security-by-design guidelines presented in *Chapter 7, Securing the Internet of Things*, are inherently covered, with full life cycle management. Many of the IoT SIM suppliers and cellular network carriers offer services with IoT SAFE that make the device security and provisioning of enterprise IoT solutions much easier. Check with your selected mobile network carrier to understand their IoT SAFE product offerings better and how this can enable a more secure enterprise IoT solution.

Over the next 10 years and beyond, the global LTE, 5G, and LPWA cellular technologies will coexist to enable new and innovative enterprise IoT solutions in many enterprise markets such as transportation, supply chain logistics, manufacturing, healthcare, and energy, as we reviewed in *Chapter 1, Transforming to an IoT Business*. Let us now expound some of the top IoT trends and new business models to provide a new perspective on enterprise IoT solutions over the next decade.

IoT trends and business models

In this section, we will review some of the top IoT trends based on enterprise IoT solutions deployed in the past 5 years and some interesting new business models enabled by these IoT solutions. Some of the leading IoT trends that will continue to grow over the next 10 years are as follows:

- Edge computing
- **Artificial Intelligence** (**AI**) and **Machine Learning** (**ML**)
- The interoperability of IoT solutions
- A social and ethical IoT
- The IoT for sustainability

Let us start with a review of edge computing which we introduced in *Chapter 6, Reviewing Cellular IoT Devices with Use Cases*.

Edge computing

Edge computing simply moves the IoT data analysis and processing in the enterprise IoT solution from the cloud to computing nodes physically closer to the device. From the perspective of the cellular network, this means moving LTE and 5G network elements closer to the end devices and customer premises to reduce network latency and improve performance. This is the concept of **Multi-Access Edge Computing (MEC)**, which we reviewed in *Chapter 6, Reviewing Cellular IoT Devices with Use Cases*. Edge computing offers several benefits to enterprise IoT solutions, including the following:

- It enables a more robust and scalable IoT solution by reducing the network load and distributing the processing to the edge, which makes the solution more resilient to single points of failure in the solution architecture

- Edge computing allows for reduced latency, power consumption, and cost by avoiding the ferrying of all IoT data to the cloud for processing

- Edge computing allows for more real-time data processing and decisions at the edge, especially when combined with IoT device AI and ML, which we will cover shortly

Some examples of IoT use cases that are driving the need for edge computing include predictive maintenance and mobile robots in smart factories; autonomous vehicles that need to collect and process real-time data about the operating environment, such as traffic and pedestrians; and video analytics used for city surveillance. In most enterprise IoT solutions, there will be a hybrid approach to edge computing where critical, low-latency processing will be implemented at the edge and other less time-sensitive IoT data used for reporting and analytics will be sent to the enterprise IoT platform. A key enabler of edge computing in the IoT device is new compute technologies for AI and ML.

AI and ML

AI is one of the hottest topics in the context of the IoT and technology in general. In the context of the IoT, AI and ML applications at the IoT device layer enable the device to interpret conditions and make decisions automatically without the need for human interaction. The most basic AI applications simply use rules or policies to trigger events or actions. ML is a more advanced form of AI where the application learns behavior rather than having it explicitly programmed. For example, predictive analysis of equipment in a smart factory is an IoT use case where ML can be used, as the normal behavior of the monitored machine is learned so that anomalous behavior can be identified and reported. Using vibration monitoring with ML in a smart factory IoT solution can identify equipment that is vibrating abnormally, indicating potential component failures and the need for the maintenance of the enterprise application. Video analytic solutions used for people recognition, traffic conditions in a smart city, and autonomous vehicles are other example use cases where AI and ML applications are creating value in IoT solutions.

Several companies have developed small ML applications that can run on constrained, low-memory, and low-power IoT devices with limited compute power. This enables new IoT solutions such as vital sign and patient monitoring in the healthcare market, inventory management in the supply chain logistics market, and crop sensors in the agriculture market. AI and ML technologies are continuing to evolve to support many IoT use cases and are key enablers for true edge computing in new enterprise IoT solutions.

As described in *Chapter 2, Understanding IoT Devices and Architectures*, and *Chapter 3, Introducing IoT Wireless Technologies*, there are several options for the architecture components in an IoT solution that will make the interoperability of devices, platforms, and network technologies a key trend in new IoT solutions.

Interoperability of IoT solutions

Interoperability can be defined as the ability of different IoT systems to work together seamlessly in exchanging and sharing data, and it is one of the central challenges in IoT. From an IoT device perspective, there are many wireless LAN protocols, including Wi-Fi, Bluetooth, and Thread, which we reviewed in *Chapter 3, Introducing IoT Wireless Technologies*, and many application protocols such as MQTT, HTTP, and CoAP, which we reviewed in *Chapter 2, Understanding IoT Devices and Architectures*. Combine this with the different IoT platforms and applications in an enterprise IoT solution and you can see the interoperability challenges within any IoT solution. A good example of this interoperability challenge is the smart home, which typically has products from several companies such as Amazon, Google, and Samsung each offering their own IoT solutions to consumers. We would like our thermostat, refrigerator, garage door opener, lights, and security cameras to work seamlessly together for a truly connected home, but each of these connected devices typically has its own siloed solution and doesn't exchange data and work together. Seamless interoperability in the smart home would enable a myriad of connectivity options where, for example, your connected car remotely turns on the thermostat and sets the home lighting when you return home, and the alarm on your smartphone automatically triggers the thermostat, lighting, and smart coffee brewer when you wake in the morning.

To address the interoperability challenge and improve IoT standardization in IoT solutions, many technology companies and device manufacturers have created alliances for adopting standards and open source development. For example, in the smart home example, Google, Amazon, Samsung, and other technology companies created the **Connectivity Standards Alliance** (**CSA**) to make IoT more accessible and usable. One of the standards from the CSA is Matter, launched in 2022, which enables smart home Wi-Fi, Bluetooth, and Thread devices to work together using a single protocol in a Matter-certified ecosystem. As shown in *Figure 10.2*, there is a myriad of standardization bodies and interoperability alliances for nearly all aspects of an IoT solution including wireless connectivity as well as transport and application protocols, with some alliances focused on vertical markets such as healthcare, connected homes, and the industrial IoT.

IoT WWAN and WLAN Radio Technologies	• 3GPP • Bluetooth Special Interest Group • DASH7 Alliance • IEEE • LoRa Alliance	• Weightless Alliance • Wi-Fi Alliance • Wi-SUN Alliance • ZigBee Alliance	
IoT Protocol Stack	• IPSO Alliance • One M2M • Open Connectivity Foundation		
IoT Verticals	**Connected Home** • Connectivity Standards Alliance • Thread Group • Z-Wave Alliance	**Industrial IoT** • Industrial Internet Consortium • Modbus • FieldComm Group	**Healthcare** • Continua Health Alliance • Allseen Alliance • Personal Connected Health Alliance

Figure 10.2 – IoT alliances and consortia

Although there is currently a lack of standardization in IoT, which contributes to the interoperability challenge, this is improving with the help of these industry alliances. In your enterprise IoT solution, it is always a best practice to leverage IoT standardization and industry alliances where possible. This is the best way to leverage the various technologies in the IoT ecosystem without *reinventing the wheel*.

With the growth in IoT solutions generating and manipulating vast amounts of data that could relate to human behavior and interaction, such as video analytics and remote patient monitoring, there will be an increased focus on social and ethical IoT solutions, which we will now discuss.

A social and ethical IoT

With IoT solutions linking systems, devices, data, and people, the IoT has the potential to create an integrated technology ecosystem that solves problems, increasing efficiency and business opportunities. The downside is the increased potential for inappropriate and unsafe behavior with unintended consequences as we saw with the advent of the internet in the early 1990s. In terms of privacy, there is a delicate balance between the use of personal data garnered from IoT solutions for safety and security and how that personal information could be used by unintended third parties for unethical purposes. For example, your health monitoring data could be shared with health insurance providers to deny health coverage, or your location data could be used by a third party to track you and your family. As the IoT grows, personal information will become more and more valuable to a diverse set of good and bad actors, including organizations, individuals, and autonomous systems. As we have seen from data breaches in many technology companies, it will be difficult to control all the personal information captured from various IoT systems. It is a delicate, complex balance in that too much privacy can lower the value of public and private IoT datasets. Overall, there is an industry-wide need to develop IoT governance guidelines for safety, security, privacy, and appropriate behavior, especially for autonomous systems. In the context of your enterprise IoT solution, following the security guidelines presented in *Chapter 7, Securing the Internet of Things*, and following the guardrails for data privacy and security in the design phase of your IoT solution are best practices to get ahead of this trend for an ethical IoT.

With the global impacts of climate change, increased energy costs and usage, and increased focus on valuable natural resources such as water, IoT solutions are at the forefront of helping address these global issues. This brings us to a review of our final IoT trend, which is the IoT for sustainability.

The IoT for sustainability

By collecting data from various sensors, including electric, gas, and water meters, as well as environmental quality sensors for air and water, IoT technologies have had a big impact on sustainability initiatives and the environment. Many companies are implementing IoT solutions for sustainability to manage energy consumption and monitor equipment to optimize operational efficiencies and reduce downtime, which also reduces operational costs. There are many use cases where IoT can support sustainable practices, including the following:

- Water, gas, and electric meters to monitor consumption and anomalies indicating issues
- Predictive maintenance on factory equipment to optimize operational efficiency
- Water leak monitoring to alert on water loss
- Environmental monitoring to identify pollutants in the air and water
- Solar and wind energy generation management for improved operational efficiencies
- Smart waste management to increase operational efficiency in waste collection and recycling
- Agricultural monitoring for irrigation, crop, and soil quality to reduce water and fertilizer usage
- Smart building monitoring to manage building power and lighting based on occupancy

These IoT use cases not only reduce enterprise operational costs but also provide a more sustainable environment for everyone. As such, sustainability initiatives will continue to be the focus of many businesses in the future. With the impacts of climate change becoming more apparent through higher temperatures, storms, and droughts, IoT solutions have the real potential to manage our natural resources better and reduce our reliance on fossil fuels in the future.

With our review of some of the top IoT trends moving forward, let us now look at some of the new business models enabled by enterprise IoT solutions that you can consider in your enterprise IoT solution.

New IoT business models

As discussed in *Chapter 1, Transforming to an IoT Business*, the IoT has enabled new recurring revenue, *as-a-service* subscription business models where the customer pays for continuous value – for example, predictive monitoring as a service or health monitoring as a service. There are also three new potential business models for enterprise IoT solutions that leverage the data, usage, and results of the solution.

These are as follows:

- Infonomics – data monetization
- An outcome-based model
- A pay-per-use model

One of the key values of an IoT solution are the insights derived from the collected data. Monetizing this data has been termed infonomics and is a growing trend within enterprise IoT solutions.

Infonomics – data monetization

Most enterprise IoT solutions are primarily designed to improve operational efficiency and provide value to the end user, but the data collected and analyzed as part of this solution could be sold to third parties given the valuable insight it provides. With this model, the IoT device or service could be offered at no cost to eliminate buying friction for the end user. In this way, many devices could be deployed to collect and analyze the valuable sensor data, which could be the energy consumption data for a building that is valuable to the building managers and utilities that need to be optimized for more efficient energy use. Another example is **User-Based Insurance** (**UBI**), which is an IoT solution from car insurance companies to monitor driving habits and offer reduced insurance premiums. The data collected provides insights into the driving behaviors of a large population of drivers and can help insurance companies offer more competitive rates. This infonomics business model can be an extension of a new or existing IoT solution implemented later in the enterprise IoT life cycle. With this model, it is important to make your customers aware of how their data will be used and safeguard it accordingly.

With most enterprise IoT solutions designed to solve problems such as reducing product service calls for equipment repairs or shortening lead times in your supply chain logistics operation, another business model to consider is the outcome-based model.

Outcome-based models

With the outcome-based model, customers pay for the outcome or benefit the IoT solution provides as opposed to the solution itself. This is appealing to most enterprise customers, as they are typically most interested in the results of the IoT solution, which could be fewer service calls, reduced factory downtime, or shorter product lead times. With this model, the "means" to the "end" for the IoT solution can be flexible, meaning the solution can be innovative in how it achieves the desired outcome. For example, this could be achieved using ML, edge compute devices connected to factory equipment, or field products to carry out predictive maintenance before a service call is needed and avoid any downtime. In this example, the solution outcome should be fewer service calls and reduced downtime, but the specific implementation of the enterprise IoT solution is mostly hidden from the customer. Where enterprise IoT solutions are expected to deliver a quantifiable outcome to an end customer, this is a good business model to consider and can reduce buying friction with customers.

In any enterprise IoT solution where you can monitor how much a customer is using the product or service, such as electric scooters or bicycles in cities, the pay-per-use model is a good business model to consider.

Pay-per-use models

With a pay-per-use model, customers are charged for the time they are actively interacting with the IoT solution product or service. In the example of electric scooters or bicycles, which are always connected to an enterprise IoT application, customers can simply use a mobile application on their smartphone to initiate service and pay for how long they use the product. Referring to the UBI example of car insurance companies, the device used to monitor driving behavior in the vehicle could also be used via a pay-per-use model to allow customers to pay insurance for the time they are driving the vehicle. When a customer's usage of an IoT solution can be monitored, the pay-per-use model may be a good option to scale your enterprise IoT solution.

In summary, there are a few key IoT technology trends, including edge computing, ML, and improved interoperability between IoT systems, which can add value and should be assessed in your enterprise IoT solution planning and design. As we discussed, with the valuable data collected in your IoT solution, it is important to consider the privacy and security of this data in an ethical IoT business model. Finally, sustainability is an important trend that should be considered to increase the long-term value of your enterprise IoT solution. In terms of new business models, beyond the standard *as-a-service* IoT models, there are three new business models we have reviewed, which can reduce or eliminate customer buying friction and enable a scalable IoT solution.

With this review of some of the key IoT trends and business models, let us now discuss how to get started on our own IoT journeys.

Getting started on your IoT journey

In this book, we provided a foundational review of IoT including the key markets, use cases, devices, wireless networks, architectures, and security guidelines with a focus on LTE and 5G IoT solutions. In *Chapter 8, Implementing an IoT Solution with Case Studies*, we presented the practical application of this knowledge for two real-world cellular IoT solution case studies. This led us to an analysis of the cellular IoT solution life cycle in *Chapter 9, Managing the Cellular IoT Solution Life Cycle*, which is the basis for getting started on your journey to implementing a successful enterprise IoT solution. As described in *Chapter 9, Managing the Cellular IoT Solution Life Cycle*, the initial design and planning stage is the most important phase to set the foundation for this successful IoT solution. To summarize, some of the key tasks in this critical stage are as follows:

- Defining the IoT business case, stakeholders, and customers
- Defining the IoT business model
- Identifying the IoT solution requirements and high-level architecture

- Determining what IoT data is needed and how it is collected, reported, stored, and secured

- Evaluating which legacy enterprise systems to integrate with the IoT solution

- Surveying cellular IoT devices, wireless carriers, and platforms that can work seamlessly with your enterprise application

- Aligning with the business leaders and the workforce on the new IoT business model

As we discussed in *Chapter 9, Managing the Cellular IoT Solution Life Cycle*, it is also important to consider each follow-up stage of the IoT solution life cycle in the initial design phase given the complexity and longevity of a typical enterprise IoT solution. Given many of the complex IoT solution architecture components such as the device, wireless connectivity, and IoT platform are outsourced to IoT ecosystem partners, one of the critical decisions in the design and planning stage is *with which IoT partner you should work to implement your IoT solution*. As shown in *Figure 10.3*, the IoT ecosystem is complex and fragmented, with various potential partners for all aspects of an enterprise IoT solution, including the device, network, cloud IoT and data analytic platforms, and system integration with a focus on various vertical IoT markets.

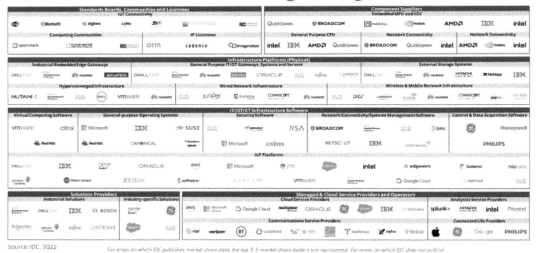

Figure 10.3 – IoT partner ecosystem

Picture credit: *IDC, "IDC Market Glance IoT and Intelligent Edge Infrastructure, 3Q22 – 2022 Sep, Doc # US49627722, Sep 2022.*

It could be overwhelming to evaluate the various partners in the IoT ecosystem as part of your enterprise IoT solution, but the taxonomy of this partner ecosystem is changing in a positive way, with many of these partners moving up the IoT architecture stack to offer additional services. For example, many cellular IoT device and module suppliers also provide connectivity, device management, and IoT platform services. Moreover, many cellular network operators also offer end-to-end IoT solution support beyond their LTE or 5G connectivity, leveraging partnerships across the IoT ecosystem. With this change to the IoT ecosystem, it is now easier to engage with fewer partners to build your IoT solution and work with those partners that gain experience developing similar enterprise solutions. Overall, it is important to engage with IoT ecosystem partners, as most businesses don't have the in-house skills to build end-to-end IoT solutions, and these partnerships allow a business to focus on the differentiated value offering of its enterprise IoT solution.

In summary, to start on your IoT journey, focus on defining the IoT business case and IoT solution requirements, working with experienced ecosystem partners in the design and planning phase of the enterprise IoT solution life cycle. Now, let's review some best practices on the road ahead of your IoT journey.

Best practices in developing and launching your IoT solution

Throughout this book, we have presented several best practices for nearly all aspects of your enterprise IoT solution including the device, wireless network, architecture, security, and life cycle management. For this final chapter on the road ahead, we will review the following three best practices in getting started on your IoT journey:

- Leveraging the IoT partner ecosystem

- Focusing on interoperability between IoT systems

- Planning on an 18-24 month launch schedule

Let's start with the IoT partner ecosystem we reviewed earlier.

Leveraging the IoT partner ecosystem

As described earlier, the IoT partner ecosystem is both extensive and fragmented with expertise in every aspect of an enterprise IoT solution, and multiple partners likely have experience implementing similar enterprise IoT solutions. Given the complexity of an IoT solution and the typical lack of in-house skills to develop and manage the full life cycle of the solution, it is crucial to leverage experienced IoT partners, especially on the device, wireless network, and IoT platform. Many system integrators have valuable experience integrating the components of your solution with legacy enterprise applications. As stated earlier, many partners offer additional IoT services beyond their traditional product offerings, which makes these partnerships more valuable by reducing interoperability issues. As a best practice, it is good to start with your selected cellular network operator and IoT platform provider to build your IoT solution partner ecosystem.

As described earlier in this chapter, interoperability between systems is one of the biggest challenges in implementing your enterprise IoT solution, so let us now look at some best practices with interoperability.

Focus on interoperability between IoT systems

As we covered earlier in this chapter, interoperability issues can manifest at many points in the enterprise IoT solution architecture, including the device interface to the monitored equipment, the device connectivity to the IoT platform, the network interface to the IoT platform, and/or the IoT platform interface to legacy enterprise applications. Resolving these interoperability issues typically requires unique skillsets most businesses do not have and takes significant time to troubleshoot and resolve. To reduce interoperability issues, it is a best practice to use standardized interfaces whenever possible. It is important again to leverage IoT partners with experience implementing similar enterprise IoT solutions and plan to spend time vetting the solution in a POC and pilot to identify any interoperability issues before the solution is fully launched. Given the extra time in resolving interop issues, it is a best practice to plan on an 18-24 month launch schedule.

Planning on an 18-24 month launch schedule

As discussed in *Chapter 9, Managing the Cellular IoT Solution Life Cycle*, following the IoT life cycle with potentially multiple POCs and pilots prior to initial solution deployment can take 18-24 months, so it is important to set the expectations of internal and external stakeholders and customers. Setting these expectations helps align the organization's workforce and senior leadership with the new enterprise IoT solution and helps ensure buy-in and the success of the solution. As noted in the best practices from the connected cooler case study in *Chapter 8, Implementing an IoT Solution with Case Studies*, completing a POC and pilot prior to launch would have reduced the number of firmware issues found in the full-scale connected cooler deployment, which created friction with stakeholders and customers.

Let us now summarize the takeaways from this chapter on the road ahead.

Summary

In this chapter, we presented some emerging cellular IoT technologies, with private networks, integrated and multi-profile SIMs, and the SIM-based IoT SAFE standard, which can improve the privacy, security, and performance of your enterprise IoT solution. We then explored the IoT trends of edge computing, AI and ML, and improvements in interoperability, which can increase the performance and value of your IoT solution. We also looked at how an ethical IoT and sustainability apply to enterprise IoT solutions. We then reviewed some new business models to reduce buyer friction with end customers. We concluded this chapter with some recommendations on how to get started and some best practices on your IoT journey.

The enterprise IoT market is strong, and we are still at the very early stages of IoT market growth, especially with new cellular technologies such as LTE LPWA and 5G enabling new and innovative use cases. Implementing an enterprise IoT solution is complex, requiring expertise in many areas of the IoT, including the device, network architecture, wireless connectivity, IoT platform, and enterprise application. The main impetus for this book has been to provide understandable, essential knowledge in these areas of the IoT, along with practical best practice recommendations to ensure the success of your enterprise IoT solution. It is our hope that this book will be a practical guide for your IoT journey into the future!

Index

Symbols

3rd Generation Partnership
Project (3GPP) 53, 73

5G
best practices 81, 82
Dynamic Spectrum Sharing (DSS) 76
frequency spectrum 76
IoT solutions 78
network architecture 74, 75
network carriers 81
network slicing 77
overview 74
private networks 81
radio technology 75
use cases 79

5G BTS 77
5G CBRS network 154-156
5G IoT devices 82
5G NR (New Radio) 73
5G NR RedCap 78, 79

A

Access Point Name (APN) 28, 54, 108, 130
Access Points (APs) 126
Advanced Encryption Standard (AES) 109
Advanced Meter Infrastructure (AMI) 36
Advanced RISC Machine (ARM) 85, 106
AI-based attacks 105
Amazon Web Services (AWS) 5, 143
American Standard for Information
Interchange (ASCII) 24
Analog-to-Digital Converter (ADC) 90, 126
application layer 22, 110
CoAP 23
LwM2M 23, 24
MQTT 23
Application Programming
Interfaces (APIs) 17
Artificial Intelligence (AI) 105, 157
asset trackers 31, 32
Attention (AT) commands 85
augmented reality (AR) 74
Automated Teller Machines (ATMs) 90

B

Base Transceiver Station (BTS) 77
BeiDou Navigation Satellite
    System (BDS) 31
best practices, IoT devices and
    architecture technologies 32
  being slow and steady 33
  established IoT technology
      partners, using 33
  off-the-shelf technology
      components, using 32
  on solution lifecycle management,
      planning 33
  security 33
Bill of Materials (BOM) 44, 85
Bluetooth 31, 32, 38
Bluetooth Special Interest Group
    (Bluetooth SIG) 38
botnets 105

C

Carrier Aggregation (CA) 75
cellular IoT device architecture 83, 84
  cellular module 85, 86
  input/output 89
  power unit 87
  processing unit 84, 85
  RF block 87
  sensors 88
  SIM 86
  WLAN radios 90, 91
cellular IoT device best practices 99
  buying, versus building 99
  carrier certification 99
  device management 100
cellular IoT device carrier certification 96

cellular IoT device types 91, 92
  alarm panel 92
  camera 92
  embedded modem 93, 94
  emergency phone 92
  hotspot 92
  in-vehicle computers 92
  lighting 93
  medical telematics 93
  Mobile Personal Emergency Response
      System (mPERS) 94
  POS device 94
  remote control device 94
  remote metering 93
  rugged handheld 94
  sensor device 94
  smart home device 95
  tracking device 95
  Vehicle OBD II 95
  vehicle telematics control unit 95
  vending telemetry device 95
  wearable 96
cellular IoT life cycle 144
  deployment 146
  designing and planning stage 144, 145
  monitoring and management 146, 147
  provisioning stage 145
  sunset process 147
cellular IoT network architecture 25, 26
  eNodeB 27
  EPC 27
  user equipment (UE) 26, 27
cellular IoT technologies
  5G CBRS network 154-156
  emerging 154
  eUICC SIM technologies 156
  IoT SAFE 157
  iUICC technologies 156
  private LTE network 154-156

cellular technology
evolution 52, 53
Cellular V2X (C-V2X) 10
Centralized Unit (CU) 77
Central Processing Unit (CPU) 84
Citizen Band Radio Service (CBRS) 154
Cloud Service Provider (CSP) 5
Complex Instruction Set
    Computer (CISC) 85
Component Carriers (CCs) 75
compound annual growth rate (CAGR) 77
Concise Binary Object Representation
    (CBOR) 24, 25
Connected Cooler business case 120
global device connectivity 120
global location, and sensor reporting to the
    enterprise cloud application 120, 121
IoT device, integrating with sensors 120
low-power operation mode,
    used for reporting 121
Connected Cooler device architecture 124
cellur module 125
input/outputs 126
power unit 126
RF 125
sensors 126
SIM (UICC) 125
WLAN radios 126
Connected Cooler network
    architecture 130-132
Connected Cooler solution 119
Connected Mode Mobility (CMM) 70
Connectivity Standards Alliance
    (CSA) 47, 159
Constrained Application Protocol
    (CoAP) 22, 23, 133
Coverage Extension (CE) mode 69
customer relationship
    management (CRM) 7

D

Datagram Transport Layer Security
    (DTLS) 23, 115, 157
data monetization 162
Demilitarized Zone (DMZ) 116
Device Provisioning Service (DPS) 145
Digital-to-Analog Converter (DAC) 90
Distributed Denial of Service
    (DDoS) attack 104
Distributed Unit (DU) 77
Downlink (DL) 59
Dynamic Spectrum Sharing (DSS) 76

E

edge computing 4, 97, 98
electronic logging device (ELD) 10
Embedded Subscriber Identity
    Module (eSIM) 86
Embedded Universal Integrated
    Circuit Card (eUICC) 59, 156
endpoint layer 106
examples 106
enhanced mobile broadband (eMBB) 73
enterprise IoT market segments 9
supply chain logistics 10, 11
transportation 10
eUICC SIM technologie 156
Evolved Node B (eNodeB) 26, 27
Evolved Packet Core (EPC) 26, 27, 74
Evolved Packet System (EPS) 57
Evolved UMTS Terrestrial Radio
    Access (E-UTRA) 54
Extended Discontinuous Reception
    (eDRX) 67, 68
best practices 68

F

Federal Communications
    Commission (FCC) **35, 96**
File Transfer Protocol (FTP) **22, 115**
Firmware Over The Air (FOTA)
    44, 66, 136, 146
Firmware over the Air/Software over
    the Air (FOTA/SOTA) **103**
Frequency Division Duplex (FDD) **60, 79**
Frequency Range 1 (FR1) **76**
Frequency Range 2 (FR2) **76**
frequency spectrum, 5G **76**

G

gateways **30**
General Authorized Access (GAA) **155**
General-Purpose Input/Output (GPIO) **84**
Global Certification Forum (GCF) **96**
Global Navigation Satellite
    System (GLONASS) **31**
Global navigation satellite
    systems (GNSS) **31**
Global Positioning System (GPS) **31, 88**
Global System for Mobile Communications
    Association (GSMA) **58, 156**
gNodeB (gNB) **75**
GPRS Tunneling Protocol (GTP) **57**

H

HaLow **39**
Heating, Ventilation, and Air
    Conditioning (HVAC) **88**
high-level IoT architecture **20, 21**
    application layer **5, 22**
    data analytics layer **5**

data format **24, 25**
device layer **4**
link layer **21**
network layer **22**
transport layer **22**
Home Network Administration
    Protocol (HNAP) **116**
Home Subscriber Server (HSS) **27, 55**
humidity sensors **88**
Hypertext Transfer Protocol (HTTP) **22, 115**

I

Industrial, Scientific, and Medical (ISM) **35**
Industry 4.0 (4th industrial revolution) **8**
Information Technology (IT) **90, 104**
input/output **89**
    digital or analog **90**
    Ethernet **90**
    Serial **90**
Integrated Circuit Card
    Identifier (ICCID) **58**
Integrated Circuits (ICs) **85**
Integrated SIM (iSIM) **59, 86**
Integrated Universal Integrated
    Circuit Card (iUICC) **59, 156**
International Mobile Equipment
    Identity (IMEI) **109**
International Mobile Subscriber
    Identity (IMSI) **27, 58, 109**
International Telecommunication
    Union (ITU) **35**
Internet of Medical Things (IoMT) **16**
Internet of Things (IoT) **101**
    business opportunities **8**
    leveraging, for digital transformation **6**
Intrusion Detection Systems/Intrusion
    Prevention Systems (IDSs/IPSs) **102**

IoT business models 157, 161
data monetization 162
outcome-based models 162
pay-per-use model 163
IoT device architectures
investigating 123, 124
IoT devices 28, 142
asset trackers 31, 32
gateways 30
remote monitoring devices 30
routers 29
serial modems 29
IoT device security, best practices 110-113
connectivity interface security 115
hardware security 113
network connectivity security 115, 116
OS security 114
software security 113, 114
Wi-Fi security 116
IoT device threats and solutions 104
AI-based attacks 105
botnets 105
IT and OT convergence 104
ransomware 105
IoT, for digital transformation 6, 7
assets and inventory, managing 7
assets and inventory, monitoring 6
business opportunities 9
customer experience, improving 9
operational efficiencies and
productivity, increasing 7, 8
smart factories, creating 7
IoT life cycle management
challenges 141, 142
IoT markets
energy/utilities 16, 17
healthcare 15, 16
industrial and manufacturing 14, 15

IoT partner ecosystem
leveraging 165
IoT platforms 143, 144
IoT protocol stack 21
IoT SAFE 157
IoT security challenges 103
IoT device threats and solutions 104
IoT security ecosystem
overview 102, 103
IoT security framework 105
application layer 110
endpoint layer 106
network layer 107-110
**IoT SIM Applet for Secure End-2-End
Communication (IoT SAFE) 156**
IoT solution
developing and launching, best
practices 165, 166
interoperability 159, 160
stage 163-165
IoT solution business cases
Connected Cooler 120
exploring 119
Smart Label solution 121
IoT solution life cycle
best practices 147
IoT solution life cycle, best practices
deployment phase 150
designing and planning stage 148
monitoring and management 151
provisioning stage 149, 150
IoT solutions key lessons 135
case studies 137, 138
device management 136
life cycle management 136, 137
solution piloting 135, 136

IoT systems
  interoperability issues, between 166
IoT technologies 4
IoT trends 157
  AI 158, 159
  edge computing 158
  for sustainability 161
  ML 158, 159
  social and ethical 160, 161
IoT use cases 44
  Bluetooth/BLE 44, 45
  LoRa 46
  LR-WPAN 46, 47
  Wi-Fi 45
IoT wireless services 142
IoT wireless technologies
  best practices 47, 48
  network topologies 37
  technical and cost comparisons 41
  WLAN 37
  WWAN 37
IP Multimedia Subsystem (IMS) 26
IPv6 over Low-Power Wireless Personal
    Area Network (6LoWPAN) 22
IT and OT convergence 104
iUICC technologie 156

J

JavaScript Object Notation (JSON) 24, 25

K

key performance indicator (KPI) metrics 32

L

licensed technologies 36
  versus unlicensed technologies 35, 36
Lightweight Machine to Machine
    (LwM2M) protocol 22-24, 133
link layer 21
Local Area Network (LAN) 90
Location-Based Services (LBS) 88, 121
Long Range Wide Area (LoRa) 21, 39, 40
long-term evolution (LTE) 36, 73, 78
LoRaWAN 39, 40
low power wide area (LPWA) 5, 30, 37, 78
Low-Rate WPAN (LR-WPAN) 21, 40, 41
LTE Category M (LTE-M) 5
LTE IoT use cases 64
LTE LPWA best practices 68
  data rate 69
  latency 69
  mobility 70
  network coverage 69
  VoLTE and SMS 70
LTE LPWA design 65, 66
LTE LPWA use-cases 70-72
LTE network 55
  APN, reviewing 57, 58
  call flow, reviewing 55-57
  LTE IoT use cases 64
  LTE radio technology 60-63
  LTE technology ecosystem 63, 64
  SIM, reviewing 58, 59
LTE radio technology 59-63
LTE technologies 54
  LTE network 54
  LTE radio technology 59
LTE technology ecosystem 63, 64

M

Machine Learning (ML) 157

Machine-to-Machine Form
    Factor (MFF2) 86

Machine to Machine (M2M) solutions 52

Machine Type Communication (MTC) 53

massive Machine Type Communication
    (mMTC) 73

massive MIMO (mMIMO) 75

Master Information Block (MIB) 56

matter 47

media access control (MAC)
    addresses 31, 126

MediaTek 85

Message Queuing Telemetry
    Transport (MQTT) 22

Microcontroller Unit (MCU) 59, 84, 157

Million Instructions Per Second (MIPS) 85

Minimum Viable Product (MVP) 150

Mobile Network Operator (MNO) 59, 156

Mobile Personal Emergency Response
    System (mPERS) 70

Mobile-Terminated (MT) data 67

Mobile Virtual Network Operator
    (MVNO) 59, 156

Mobility Management Entity (MME) 27

module 85

MQTT for Sensor Networks (MQTT-SN) 23

Multi-Access Edge Computing
    (MEC) 98, 158

Multiple Input Multiple Output
    (MIMO) 59, 75

Multiprotocol Label Switching (MPLS) 109

N

Narrow Band IoT (NB-IoT) 5

Network Address Translation (NAT) 108

network architectures
    analyzing 129
    of Connected Cooler 130-132
    of Smart Label 132

network capacity 43

network carriers, 5G 81

network functions (NFs) 77

network layer 22, 107-110
    IP blacklists/whitelists 108
    non-routable IP addresses 108
    peer-to-peer blocking 108
    private IP pools 108

network slice selection function (NSSF) 77

network slicing, 5G 77

network topologies 37

Non-IP Data Delivery (NIDD) 27

Non-Recurring Engineering (NRE) 142

Non Standalone (NSA) architecture 74, 154

O

On-Board Diagnostic II (OBD II) 95

Operational Technology (OT) 104

Original Device Manufacturers (ODMs) 142

Original Equipment Manufacturers
    (OEMs) 142

Orthogonal Frequency Division
    Multiplexing (OFDM) 75

outcome-based models 162

over-the-air (OTA) 48

P

Packet Data Network Gateway (PGW) 55

Packet Data Network (PDN) 27, 57

Packet Data Protocol (PDP) 27

pay-per-use model 163

PCS Type Certification Review
    Board (PTCRB) 96

Point-of-Sale (POS) 90

point-to-point (P2P) wireless 45

Policy and Charging Rules
    Function (PCRF) 28

Power Saving Mode (PSM) 68, 67
  best practices 68

power unit 87

Pre-Shared Key (PSK) 116

pressure sensors 89

Printed Circuit Board (PCB) 44, 86, 128

Priority Access Licenses (PALs) 155

private 5G networks 81

private LTE network 154-156

Programmable Logic Controllers (PLCs) 90

proof of concept (POC) 33, 146

Public Secure Packet Forwarding
    (PSPF) 116

Pyroelectric Infrared (PIR) sensor 89

Q

Qualcomm 85

Quality of Service (QoS) 28, 155

R

Radio Access Network (RAN) 27,
    54, 75, 98, 108, 130

radio frequency (RF) 31, 36

Radio Resource Connection (RRC) 56

Radio Resource Control (RRC) 67

radio technology, 5G 75

Random Access Memory (RAM) 84

Random Access Preamble (RAP) 56

ransomware 105

Read-Only Memory (ROM) 84

Real-Time Operating System (RTOS) 85

received signal strength indicator (RSSI) 31

Registered Jack-45 (RJ-45) 90

Release Assistance Indicator (RAI) 68

remote monitoring devices 30

remote patient monitoring (RPM) 4, 45

Representational State Transfer
    Application Programming
    Interfaces (REST APIs) 132

Rest of World (RoW) 125

Return on Investment (ROI) 155

router 29

S

Secure Shell (SSH) 116

sensors 88
  accelerometer 88
  gas detection 89
  GPS 88
  proximity 89
  temperature 88

serial modem 29

service-based architecture (SBA) 74

Service Capability Exposure
    Function (SCEF) 28

Service Level Agreements (SLAs) 33

Serving Gateway and PDN Gateway
    (S-GW/P-GW) 28

Short Message Service (SMS) 26, 67

Signaling Gateway (SGW) 26
Signaling Radio Bearer (SRB) 57
Simple Mail Transfer Protocol (SMTP) 115
Smart Grid 17
Smart Label architecture 132, 133
Smart Label device architecture 127
  cellular module 127
  power unit 128, 129
  RF 128
  sensors 128
  SIM (iUICC) 128
Smart Label solution business case 121, 122
  global connectivity 122
  global location and sensor reporting 122
  low-cost for one-time use 123
  low-power operation, with battery 123
Software Over The Air (SOTA) 146
Spectrum Access System (SAS) 155
Standalone (SA) architecture 74, 154
sub-6 GHz 76
Subscriber Identity Module
    (SIM) 26, 54, 108
Subscription Management Data
    Preparation (SM-DP) 59
Subscription Management Secure
    Routing (SM-SR) 59
Subscription Management (SM) 59
Supervisory Control and Data
    Acquisition (SCADA) 29, 90
supply chain logistics, IoT markets 10-12
  asset monitoring 13
  inventory/warehouse management 11
  transportation and fleet management 13, 14
System Information Block (SIB) 56
System on Chip (SoC) 59

T

Tamper-Resistant Element (TRE) 59, 156
technical and cost comparisons,
    IoT technologies 41
  data throughput 44
  frequency spectrum 42
  infrastructure cost and availability 42
  interference 42
  module cost 44
  network capacity 43
  power consumption 43
  range 43
Telematics Control Unit (TCU) 95
Teletype Network (TELNET) 115
temperature, sensors
  humidity sensors 88
  pressure sensors 89
Time Division Duplex (TDD) 60
Total Radiated Power (TRP) 87
Tracking Area Update (TAU) 67
Transmission Control Protocol (TCP) 22
transport layer 22
Transport Layer Security
    (TLS) 115, 130, 157

U

Ultra Reliable Low Latency
    Communications (URLLC) 73
Universal Integrated Circuit
    Card (UICC) 58
Universal Mobile Telecommunication
    System (UMTS) 54
Universal Plug and Play (uPnP) 116
Universal Time Coordinated
    (UTC) format 134

unlicensed spectrum bands, IoT solutions
   2.4GHz 36
   5.8GHz 36
   900 MHz 36
unlicensed technologies
   versus licensed technologies 35
use cases, 5G 79
   Industrial automation (Industry 4.0) 80
   remote surgery 80
   smart vehicles (autonomous
     vehicles and V2X) 80
   VR/AR 80
User-Based Insurance (UBI) 95, 162
User Datagram Protocol (UDP) 22
user equipment (UE) 26, 75

V

Virtual Private Network (VPN) 28, 107
virtual reality (VR) 74
Voice over IP (VOIP) 44
Voice over LTE (VoLTE) 26, 66, 92

W

Wide Area Network (WAN) 90
Wi-Fi access point (AP) 31, 43
WiFi Protected Access 2 (WPA2) 116
Wireless Fidelity (Wi-Fi) 21
Wireless Local Area Network
   (WLAN) 4, 29, 52
Wireless Personal Area Networks
   (WPANs) 37
Wireless Wide Area Network (WWAN) 5, 29
WLAN radios 90, 91
WLAN/WPAN unlicensed technologies
   Bluetooth/BLE 38
   LoRa 39, 40
   LoRaWAN 39, 40
   LR-WPAN (IEEE 802.15.4) 40, 41
   Wi-Fi 38, 39

`Packt.com`

Subscribe to our online digital library for full access to over 7,000 books and videos, as well as industry leading tools to help you plan your personal development and advance your career. For more information, please visit our website.

Why subscribe?

- Spend less time learning and more time coding with practical eBooks and Videos from over 4,000 industry professionals

- Improve your learning with Skill Plans built especially for you

- Get a free eBook or video every month

- Fully searchable for easy access to vital information

- Copy and paste, print, and bookmark content

Did you know that Packt offers eBook versions of every book published, with PDF and ePub files available? You can upgrade to the eBook version at `packt.com` and as a print book customer, you are entitled to a discount on the eBook copy. Get in touch with us at `customercare@packtpub.com` for more details.

At `www.packt.com`, you can also read a collection of free technical articles, sign up for a range of free newsletters, and receive exclusive discounts and offers on Packt books and eBooks.

Other Books You May Enjoy

If you enjoyed this book, you may be interested in these other books by Packt:

Building IoT Visualizations using Grafana

Rodrigo Juan Hernández

ISBN: 978-1-80323-612-4

- Install and configure Grafana in different types of environments
- Enable communication between your IoT devices using different protocols
- Build data sources by ingesting data from IoT devices
- Gather data from Grafana using different types of data sources
- Build actionable insights using plugins and analytics
- Deliver notifications across several communication channels
- Integrate Grafana with other platforms

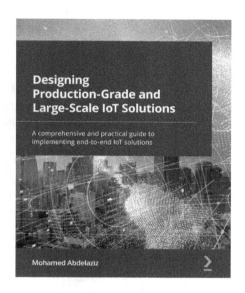

Designing Production-Grade and Large-Scale IoT Solutions

Mohamed Abdelaziz

ISBN: 978-1-83882-925-4

- Understand the detailed anatomy of IoT solutions and explore their building blocks
- Explore IoT connectivity options and protocols used in designing IoT solutions
- Understand the value of IoT platforms in building IoT solutions
- Explore real-time operating systems used in microcontrollers
- Automate device administration tasks with IoT device management
- Master different architecture paradigms and decisions in IoT solutions
- Build and gain insights from IoT analytics solutions
- Get an overview of IoT solution operational excellence pillars

Packt is searching for authors like you

If you're interested in becoming an author for Packt, please visit `authors.packtpub.com` and apply today. We have worked with thousands of developers and tech professionals, just like you, to help them share their insight with the global tech community. You can make a general application, apply for a specific hot topic that we are recruiting an author for, or submit your own idea.

Share Your Thoughts

Now you've finished *Implementing Cellular IoT Solutions for Digital Transformation*, we'd love to hear your thoughts! Scan the QR code below to go straight to the Amazon review page for this book and share your feedback or leave a review on the site that you purchased it from.

`https://packt.link/r/180461615X`

Your review is important to us and the tech community and will help us make sure we're delivering excellent quality content.

Download a free PDF copy of this book

Thanks for purchasing this book!

Do you like to read on the go but are unable to carry your print books everywhere? Is your eBook purchase not compatible with the device of your choice?

Don't worry, now with every Packt book you get a DRM-free PDF version of that book at no cost.

Read anywhere, any place, on any device. Search, copy, and paste code from your favorite technical books directly into your application.

The perks don't stop there, you can get exclusive access to discounts, newsletters, and great free content in your inbox daily

Follow these simple steps to get the benefits:

1. Scan the QR code or visit the link below

https://packt.link/free-ebook/9781804616154

2. Submit your proof of purchase
3. That's it! We'll send your free PDF and other benefits to your email directly

www.ingramcontent.com/pod-product-compliance
Lightning Source LLC
Chambersburg PA
CBHW060600060326
40690CB00017B/3780